D0174893

CONCRETE PROBLEMS

CONCRETE PROBLEMS

Causes and Cures

JOHN C. ROPKE

Concrete Consultant

McGRAW-HILL BOOK COMPANY

New York St. Louis San Francisco Auckland
Bogotá Hamburg Johannesburg London Madrid Mexico
Montreal New Delhi Panama Paris São Paulo
Singapore Sydney Tokyo Toronto

To Elaine, who has always been more than a wife, and to our sons, John and Richard.

Library of Congress Cataloging in Publication Data

Ropke, John C..
Concrete problems.

Includes index.

1. Concrete. I. Title.
TA439.R73 666'.893 80-28433
AACR1

1 2 3 4 5 6 7 8 9 0 KPKP 8 9 8 7 6 5 4 3 2 1

ISBN 0-07-053609-0

The editors for this book were Joan Zseleczky and Celia Knight, the designer was Elliot Epstein, and the production supervisor was Teresa F. Leaden. It was set in Baskerville by The Kingsport Press.

Printed and bound by The Kingsport Press.

CONTENTS

Preface

Of all the materials used in construction, concrete appears to be the most complex and least understood. However, the excellent features of concrete are undeniable. It is readily available; it is strong, resists fire, and will not rot; and it requires little or no maintenance when properly handled. It is sturdy under storm forces, resists rodent and insect attack, and is the very foundation of the construction industry.

Thousands of cubic yards of concrete can be placed without incident, and yet, on any given day, problems may develop for no *apparent* reason. Obviously there has to be something different on the problem day than on the previous trouble-free days of the placement.

With all of the advantages in concrete's favor, the occasional problem is a small price to pay for the many benefits. With a greater knowledge and understanding of the causes that contribute to and create problems, most problems can be eliminated.

It is hoped the contents of this book will help in understanding the theory of concrete, its limitations, the factors that may cause problems and how to avoid them, and how to correct concrete problems should they occur.

Over the past 20 years I have conducted concrete seminars for architects, engineers, state highway personnel, contractors, concrete technicians, and students. Records have been kept of the most frequently asked audience questions, and this manuscript was prepared with these questions in mind.

John C. Ropke

NOTE: In this book masculine nouns and pronouns are sometimes used to simplify language. These masculine references apply to women as well as to men, for they are intended in a purely generic sense.

ACKNOWLEDGMENTS

Special gratitude is due Mr. Sandy Herod, editor of Pit and Quarry Publications, Chicago, Illinois, for his help and encouragement, and Mr. Robert Lewis of Marist College, Poughkeepsie, New York.

My thanks to the many people over the years who have given readily of their knowledge and experience, particularly the U.S. Navy, the chemists and engineers of the Johns-Manville Corporation Research Laboratory, Manville, New Jersey, and to the following:

Publications: *Concrete Construction* (Periodical), Concrete Construction Publications, Inc., 329 Interstate Road, Addison, IL 60101

NAWIC Construction Dictionary, 4th ed., NAWIC (National Association of Women in Construction), P.O. Box 6142, Phoenix, AZ 85005

Modern Concrete (periodical), 105 West Adams Street, Chicago, IL 60603

Associations: American Concrete Institute (ACI), Box 19150, Detroit MI 48219

American Society for Testing and Materials (ASTM), 1916 Race Street, Philadelphia, PA 19103

National Ready Mixed Concrete Association (NRMCA), 900 Spring Street, Silver Spring, MD 20910

Companies and Other Businesses: Alpha Portland Cement Company, 15 South Third Street, Easton, Pa.

Andren Concrete Company, Dover Plains, N.Y.

Universal Atlas Cement Company, Subsidiary of Heidelberger Zement AG; formerly a subsidiary of United States Steel

Fairway Testing Laboratory, Stony Point, N.Y.

J. C. R.

UNIT CONVERSION TABLE

U.S. Customary	*Metric*	*Imperial*
1 pound (lb)	0.454 kilogram (kg)	1 pound
1 gallon (gal)	3.758 liters (L)	0.833 Imperial gallon (Imp gal)
1 inch (in)	2.54 centimeter (cm)	1 inch
1 pound per square inch (psi)	0.0703 kilogram per square centimeter (kg/cm^2)	1 pound per square inch
1 cubic yard (cu yd)	0.7646 cubic meter (m^3)	1 cubic yard

CONCRETE PROBLEMS

1/CONCRETE MATERIALS

CEMENT: HISTORY

The most famous use of cementitious materials dates back more than 23 centuries to the Roman construction of aqueducts and the Appian Way. Very little was known of these early cementing materials until 1756, when John Smeaton, an English government employee, discovered that a mixture of limestone and clay would harden into a solid mass when burned.

A patent was obtained in 1796 by a James Parker of England for his so-called Roman cement, which later became known as *natural cement.* Parker's process consisted of burning stone or clay products in a limekiln and grinding the clinkers into a powder.

In 1824, Joseph Aspdin, an English stonemason, took out a patent for an improved cement which he called *portland cement,* so named because appearance of the hardened cement resembled a very hard gray stone found on the Isle of Portland, England. Aspdin appears to have discovered that a more intense burning of the limestone and clay mixture produced a harder cement than did the earlier process.

Experiments continued to yield minor improvements on Aspdin's manufacturing process for some 20 years, until Isaac Johnson discovered a method of grinding that pulverized the small pieces of clinker which the earlier process failed to powder. Over the following years additional modifications were made. Chemical compounds were added, and various burning and grinding methods were experimented with and improved upon.

Manufacturing Process

Modern portland cement is a finely pulverized material consisting of four main ingredients, namely, lime, silica, alumina, and iron oxides.

Portland cement is varied in type by changing the percentages of these four main chemical compounds and by adjusting the degree of fineness. These materials are fed into a long steel cylindrical kiln, known as the *rotary kiln,* for calcination at temperatures of about 2700°F.

The clinker produced is then cooled and pulverized. During this process a small amount of gypsum is added to retard the time of set, as the original ingredients would otherwise set too rapidly for practical use. Included in the pulverizing process are chemical grinding aids, added in small amounts, to increase fines and promote increased production.

Portland Cement Types

The standard portland cements are (1) portland cement Types I through V, (2) portland blast-furnace slag cement, and (3) Portland-pozzolan cement. Types I through V are discussed in detail.

Type I A general-purpose cement used in any construction where special properties of other types are not required.

Type II A modified cement used in general construction where the concrete is exposed to moderate sulfate attack. It is a slightly retarding cement that generates heat at a slower rate. Because it delays set somewhat, it is a good choice for concrete placed in warm weather.

Many concrete producers switch cement types from the use of Type I in the colder months to Type II when normal ambient temperatures are higher. Type II is also the logical selection for traffic areas subjected to heavy applications of deicing chemicals or salts.

Type III A high- and early-strength portland cement used when higher strengths are desired at very early ages. It permits earlier removal of forms and allows passage of traffic at an earlier age.

At the age of 3 months, Types I, II, and III are approximately equal in strength. Because of infrequent demand, many concrete producers do not store Type III cement in their cement silos. An extra bag of Type I or II cement per cubic yard of concrete, mixed at high speed at the jobsite before unloading, will create a concrete that meets higher early-strength requirements. This substitution is usually acceptable to public works departments as well as to general contractors.

Type IV A low-heat-of-hydration cement used in massive structures, such as dams, to reduce heat of hydration.

Type V A high-sulfate-resistant cement used to protect concrete exposed to severe sulfate action, as in areas where soils or nearby waters have a high alkali content.

Some of the cements mentioned above are manufactured with air-entraining agents included. Although popular years ago, air-entrained cements are presently in less demand because it is often impossible to maintain a reasonably consistent air content in the concrete when using them under certain conditions. Everything else being equal, as the concrete temperature rises, air contents decline. However, in warm weather there is little a concrete producer can do to satisfactorily control concrete temperatures when using air-entrained cement.

With non-air-entrained cement, where the liquid air-entraining agent is added at the concrete plant through a calibrated dispenser, the producer can increase or decrease the dosage of the agent to comply with the required air content as conditions change.

Many different kinds of cement other than the portland cement used in everyday concrete are manufactured for a variety of purposes. Mortar cement, for example, is not a portland cement. It is manufactured differently to produce a desired characteristic for the purpose of laying up masonry units with better adhesion, and it is more or less self-curing. Keene's and magnesite cements are not portlands. Some cements are modified portlands, such as nonshrink cement, expansive cement, oil well cement, aluminous cement, and block and pipe cement.

In addition to the standard gray portland cement, many other colored cements are being manufactured to fill a growing need. In recent years architects have been trying to get away from the traditional concrete look by specifying various colored exposed aggregates with matching, or contrasting, colored cements.

Cement Stains

White cement is usually available in most areas. When a special colored cement is required, color pigments can be added to the mortar or concrete mix to obtain the tone desired. Further, *cement* stains are available that can be sprayed, rolled, or brushed on hardened concrete surfaces with good to excellent effect. One should be sure, however, that the stain chosen will not fade in the sun.

WATER: POTABILITY

The accepted criterion for concrete-mixing water is that it be drinkable. There is strong evidence, however, that some treated waters may not act favorably with all admixtures or with different brands of cement.

Recently it has been discovered that the water may well be the source of higher field-strength results than expected. It is not unusual to have concrete cylinders cast in the field with higher strengths than cylinders of the same mix with equal slump from a testing laboratory. Certainly, the mix designed under controlled conditions in the laboratory should normally be expected to produce higher-strength cylinders than cylinders cast on the jobsite. It is for this reason that specifications require that laboratory strengths be designed for at least 15 percent higher than the strength required on the jobsite.

If the brands of cement, aggregates, and admixtures used in the field were the same as in the laboratory design mix, why is there a difference? Many possibilities can be offered for lower field-strength results, but higher strengths are more difficult to explain.

Water for Test Cylinders

There is evidence that indicates that the water, in some cases, is the answer to the higher than expected field-test results. When samples of the cement, aggregates, and admixtures are sent to a testing laboratory for the determination of mix designs and of the water/cement (W/C) ratio, a sample of the water used at the concrete plant should be included, unless it is known that both parties use water from the same source.

FINE AGGREGATE: IMPORTANCE

Most concrete specifications include a requirement for quality fine aggregate to ensure the quality of the hardened concrete. Although poor-quality natural sand may cause no deterioration under freeze/thaw conditions in air-entrained concrete, it causes significant variation in mortar shrinkage and water demand.

Fine aggregate is used in concrete to improve the properties of the plastic mix, facilitate finishing, promote uniformity, and inhibit segregation. These improvements are accomplished largely by the grading, size, shape, and surface texture of the particles.

Except for lightweight aggregate, fine aggregate for concrete must consist of natural sand, manufactured sand, or a combination of both.

Sand Types

ASTM (American Society for Testing and Materials) defines *natural sand* as a fine granular material resulting from the natural disintegration of rock or from the crushing of friable sandstone.

Manufactured sand is the fine material resulting from the crushing and classification (by screening or other means) of rock, gravel, or blast-furnace slag.

Sand must meet the standard tests for soundness, organic impurities, and deleterious materials that may react unfavorably with the alkalies in the cement; it must be graded within specified limits. The standard sieves for grading fine aggregate are No. 4, No. 8, No. 16, No. 30, No. 50, and No. 100.

Gradation Requirements

A typical test of fine aggregate is shown in Table 1-1.

Table 1-1

Sieve size	*Aggregate retained, g*	*Aggregate retained, %*	*Aggregate passing, %*	*Specification limits, %*
No. 4	0	0	100	95–100
No. 8	80	14	86	80–100
No. 16	207	36	64	50–85
No. 30	320	56	44	25–60
No. 50	438	76	24	10–30
No. 100	533	93	7	2–10
Pan	574			

$$\text{FM: } \frac{275}{100} = 2.75$$

Fineness Modulus

The *FM* (*fineness modulus*) is an indication approximately proportional to the average particle size of the tested aggregate. The lower the fineness modulus, the finer the sand. If the sand meets the gradation specifications, the FM is a reliable indication of its performance. However, it does not point out the difference between a gap-graded material and an ideal one.

Coarse Sand: Problems, Cures

Ideally graded sand is difficult to obtain in many areas. A coarse fine aggregate (FM over 3.00) is reasonably acceptable in very rich mixes, but in the standard leaner mixes delivered in everyday concrete, it is harsh under the trowel, causes segregation, and creates unsightly sand streaking in walls.

If fine mason sand is available, it may be blended with the coarse concrete sand to alleviate much of this problem. Where no fine sand is available, the use of an air-entraining agent along with a water-reducing agent introduced into the mix will greatly improve the acceptability of the mix.

Fine aggregate purchased from a source approved by state or local agencies assures the buyer that the material is sound. Even then, however, caution is necessary. The gradation, particle size, shape, and surface texture are not necessarily approved. It becomes the duty of the buyer to assure himself of these important requirements for use in quality concrete.

The gradation is of particular importance. An ideal gradation would occupy a numerical position somewhere in the center of the specification limits. A fine aggregate might show, for example, 84 percent passing the No. 16 sieve and 25 percent passing the No. 30 sieve. Although within allowable limits, such aggregate would create a gap gradation and result in unsatisfactory concrete. It would segregate more readily, bleed more profusely, and present finishing difficulties.

Nature did not leave ideally graded fine aggregate at the site of the sandbank. Close attention to the ever-changing source of the material and tight quality control of the finished product are required to maintain reasonably consistent gradation. In spite of the best of intentions of the supplier, however, there can be occasions when the aggregate delivered to the concrete manufacturer is faulty.

It becomes the duty, then, of the concrete producer to maintain quality control by conducting daily gradation tests of the purchased product. Even under normal conditions the FM is subject to minor variations. A record of daily sand tests, which will show the average FM over an extended period of time, is therefore necessary for designing concrete mixes when using the FM formula.

Although concrete mixes containing coarse fine aggregate with an FM over 3.00 may produce reasonably acceptable concrete under average conditions, they frequently present problems when used for pumping. The introduction of air-entraining agents and plasticizers may well help solve the problem temporarily, but the eventual solution is to obtain fine aggregate with a lower FM.

Impurities: Tests

It is necessary that sand be clean in order to ensure quality concrete. The impurities that occur in sand are loam, clay, mica, and organic matter. Sand containing more than 3 percent of loam or clay should not be used.

A 1-qt mason jar can be used to check the silt or loam content. Fill the jar with 2 in. of the dry sand to be tested and add water until the container is about three-quarters full. Cap and shake vigorously, ending with a gentle swirling motion to level the sample.

The water above the sample will be murky, clouded by the fine particles held in suspension. Let the sample stand overnight. Gradually these fines will settle at the top of the sample, showing a layer of buff-colored material readily distinguishable from the rest of the sample. If this top layer measures over ⅛ in., the sand is objectionable for use in concrete and must be washed more thoroughly before acceptance.

Organic matter is sewage, vegetable matter, tannic acid, manure, and similar substances of this nature. In excessive amounts any of the above is objectionable.

To test for organic matter, use a 12 fl oz clear-glass bottle available at any drugstore. Fill the bottle to the 4½-oz mark with the sand to be tested. Add to this a 3% solution of sodium hydroxide in water until the volume of the sand and liquid, indicated after shaking, is 7 fl oz. (The caustic solution can be made by adding 1 oz of sodium hydroxide to 1 qt of water.) Stopper the bottle, shake vigorously, and allow it to stand for 24 hours. If there are no organic substances present, the liquid above the sample will be clear or nearly so. Slightly straw-colored liquid is not objectionable. Beyond this stage the liquid begins to take on the color of tea, light to very dark. A light, weak tea color is the border line for acceptable use of the tested sand. Shades darker than this indicate higher-than-acceptable percentages of organic matter in the sand, and such sand should not be used in concrete.

COARSE AGGREGATE: KINDS

The three coarse aggregates used in standard concrete mixes are

1. Gravel and crushed gravel
2. Crushed stone
3. Air-cooled blast-furnace slag

Particle Shape, Size, Soundness

The particle shape and surface texture of aggregates influence their performance in concrete. Aggregate containing flats or elongated particles will reduce workability. Concrete made with such materials requires higher sand contents, necessitating more water for a given slump. Both these additions in turn demand a higher cement factor to compensate for the increased surface area of the added sand and the higher water content if strength is to be maintained.

In freeze/thaw areas the porosity and absorption factor of the coarse aggregate is of particular importance. Water expands roughly 9 percent when it freezes. If the aggregate particles absorb more water than their pore structure can accommodate, popouts occur on the concrete surface, and cracking or even partial failure may result.

If a perfectly sound aggregate is not readily available in a concrete producer's area, reducing the maximum size of the aggregate will be beneficial. Avoid sizes over 1 in.; a ¾-in. graded coarse aggregate is ideal. However, a well-graded 1-in. aggregate requires less cement, sand, and water for a given slump and strength than a ¾-in. material. This factor should be considered when reducing the coarse aggregate size.

The gradation of the coarse aggregate, as well as that of the sand, is an important factor in the workability of the plastic concrete and the ultimate strength of the hardened mass. Although strength alone is not a guarantee of durability in exposed concrete, a sound, well-graded coarse aggregate is good insurance. It is important that the aggregate resist weathering, and that there is no unfavorable reaction between the aggregate minerals and the components of the cement. If there is, the concrete, in time, will undergo partial or complete disintegration.

Gradation Requirements

The two aggregate sizes used in ready-mix concrete are generally No. 57 and No. 67. It will be noted that the range of permissible percentages passing both these sizes is quite large: 25 to 60 percent pass on No. 57 for the ½-in. sieve, and 20 to 55 percent pass on No. 67 for the ⅜-in. sieve. Although a tested aggregate may just meet these specifications on the high or low side, it does not guarantee a good workable concrete with maximum strength for a given cement factor. The middle of the allowable range might seem desirable, but the particle shape and size of the aggregate may require a higher or lower percentage of material passing for best results.

Aggregates from an approved source, or those from a material manufacturer with a history of proven quality, provide a reasonable guarantee of soundness. However, the gradation of the material, or particle shape and size, can vary with each different supplier. Time and the elements change the nature of most sand and gravel banks. The face of most deposits is not the same throughout the year. But with good quality control the finished product will remain reasonably consistent.

Deviations

Minor deviations from the norm can be expected from most suppliers. Generally these deviations in gradation will not seriously affect the concrete mix unfavorably. It is wise, however, to keep a check on the passing

Size number	Nominal size (sieves with square openings)	Amounts finer than each laboratory sieve (square openings), wt %												
		4 in. (101.6 mm)	3½ in. (90.5 mm)	3 in (76.1 mm)	2½ in. (64.0 mm)	2 in. (50.8 mm)	1½ in. (38.1 mm)	1 in. (25.4 mm)	¾ in. (19.0 mm)	½ in. (12.7 mm)	⅜ in. (9.51 mm)	No. 4(4.76 mm)	No. 8(2.38 mm)	No. 16(1.19 mm)
1	3½ to 1½ in. (90.5 to 38.1 mm)	100	90–100		25–60		0–15		0–5					
2	2½ to 1½ in. (64.0 to 38.1 mm)			100	90–100	35–70	0–15		0–5					
357	2 in. to No. 4 (50.8 to 4.76 mm)				100	95–100		35–70		10–30		0–5		
467	1½ in. to No. 4 (38.1 to 4.76 mm)					100	95–100		35–70		10–30	0–5		
57	1 in. to No. 4 (25.4 to 4.76 mm)						100	95–100		25–60		0–10	0–5	
67	¾ in. to No. 4 (19.0 to 4.76 mm)							100	90–100		20–55	0–10	0–5	
7	½ in. to No. 4 (12.7 to 4.76 mm)								100	90–100	40–70	0–15	0–5	
8	⅜ in. to No. 8 (9.51 to 2.38 mm)									100	85–100	10–30	0–10	0–5
3	2 to 1 in. (50.8 to 25.4 mm)				100	90–100	35–70	0–15		0–5				
4	1½ to ¾ in. (38.1 to 19.0 mm)					100	90–100	20–55	0–15		0–5			

Figure 1-1 Grading requirements for coarse aggregate. (Source: American Society for Testing and Materials, Specification C 33, "Specification for Concrete Aggregates.")

percentage of the ⅜-in. size (pea gravel). Many masons prefer this size for concrete, and this size is used in road construction.

When there is a high demand for pea gravel from the aggregate producer, there could well be a decrease in that size from the norm on the larger-graded concrete material. Conversely, if there is little demand, more of the ⅜-in. size than usual may be added to the product delivered to the concrete producer.

Either change, if drastic, even when within specification limits, can change the workability, pumpability, finishing, and strength of concrete. It is difficult to know when these changes take place unless the concrete producer runs daily gradation tests at his plant and keeps records.

Clean uncrushed gravel or crushed stone with equal gradation will produce equal strengths with the same cement factor. However, crushed aggregates tend to segregate less than rounded aggregates (1) when stockpiled in a cone shape and (2) when being removed from the storage hopper in a concrete plant.

No matter how ideally graded the coarse aggregate is at the manufacturer's pit or quarry, it is almost impossible to maintain this gradation until it enters the concrete delivery truck. Some minor segregation will occur during the loading of the aggregate delivery truck and during the trip to the concrete plant. Further segregation will occur when the aggregate is unloaded at the concrete producer's plant. Where plant space is limited, the aggregate is often stockpiled by crane in a high cone-shaped pile, creating serious segregation, particularly when the coarse aggregate is round and has a high percentage of large particles.

When possible, regardless of the shape or maximum size of the coarse aggregate, it is preferable to dump this delivered material directly into the plant receiving bin. Here again, as the aggregate is drawn from the bin and emptied into the weigh scale below, it will segregate. Aggregate does not level itself as water would, but becomes an inverted cone, creating the segregation. Keeping the bin reasonably full will help cure the problem. In view of these facts, it cannot be overemphasized that one should purchase the best-graded coarse aggregate available.

When considering the problems above, samples taken for tests should be obtained from the discharge gate at the bottom of the holding bin above the weigh scale. Take a sample of at least 25 lb. If there is any doubt about the first test, run another from a different sample.

A few years ago a concrete supplier had difficulty obtaining crushed stone with a sufficient percentage of ⅜-in. material. A local gravel manufacturer had a stockpile of ¾-in. material with an excess of ⅜-in. material. Mix designs were made of approximately 50 percent crushed stone and 50 percent of this gravel with excellent results.

Lightweight Aggregate Absorption

Lightweight structural coarse aggregate is becoming more in demand each year. The specific gravity of this material is lower than that of standard concrete aggregates, reducing the unit weight of the concrete (the weight of 1 cu ft) from 145–150 lb for standard-weight concrete to 115–120 lb for the lightweight concrete.

The absorption factor of this type of aggregate varies greatly, depending largely on the type of material and the manufacturing process. Almost complete saturation of lightweight aggregate is necessary for consistency in the field. Saturation is a must when the lightweight concrete is to be pumped. If the aggregate is only partially wetted, it will absorb a portion of the mixing water, causing loss of slump, workability, and finishability.

On the jobsite, if water is added frequently during the discharging of the concrete to maintain slump, the strength will be reduced. In some cases, where the aggregate continues to absorb water during the finishing process, serious cracking will result. Stockpiling the aggregate in small piles close together and sprinkling water from a revolving-type lawn sprinkler for 72 hours before use will usually ensure adequate moisture. Remember to take this moisture into account when batching.

Unit Weight, Voids, Absorption

The unit weight of a material in concrete is the weight of 1 cu ft of that material. This unit weight would include the normal voids (spaces between the aggregate particles) of the material.

All aggregates are delivered in a damp or wet condition. The percentage of wetness is referred to as the *moisture content.* The smaller the aggregate particle size, the more moisture it is likely to contain. In sand, for example, the small size of the individual particles creates a greater overall surface area and a higher number of voids to entrap moisture than in the case of the larger coarse aggregates.

Fine aggregate will normally range in moisture content from 1 percent to as high as 10 percent, while coarse aggregate will normally not contain more than 2 percent. This moisture content refers only to free water and has little to do with the porosity or absorption factor of the individual particles of the aggregate.

If we were to wipe dry all the individual particles, the material would then be referred to as being in an *SSD* (*saturated surface dry*) condition. If we were to place a preweighed sample of this SSD material in an oven for 24 hours to dry it out internally, the aggregate would then be considered as being in a *bone-dry* condition. The difference between both samples would be the percentage of absorption.

In the design of concrete mixes the aggregate weights are figured as SSD weights. As most aggregates used in the production of concrete are never in a bone-dry condition, the total water content per cubic yard of concrete for a given slump is constant, disregarding the absorption factor.

Average concrete sand has an absorption range of between 0 and 2 percent by weight. Average coarse aggregate will generally not exceed 1 percent. Concrete aggregates with high porosity and absorption factor create internal stresses reducing concrete durability.

2 / ADMIXTURES

Admixtures are materials other than cement, aggregates, and water that are added to concrete. They are used primarily to produce a given quality in, or to improve the normal qualities of, nonadmixture concrete.

For many years good quality concrete was produced without the use of admixtures. Strong, durable, workable, finishable concrete made with proper mix design and suitable materials proved satisfactory. Specifications are still being written that permit air-entraining agents but no other admixture in the concrete.

In general, however, it is agreed by many in the industry that the use of suitable admixtures can be economical and impart otherwise unobtainable characteristics to concrete. The admixture chosen should, of course, meet all standard specifications for its type.

AIR-ENTRAINING AGENTS: QUALITY FEATURES

The introduction and acceptance of this important type of admixture opened the door for others to come. Although it causes a slight loss of strength in concrete mixes, the benefits of an air-entraining agent are legion.

Air entrainment, discovered in the mid-nineteen-thirties, was principally used in concrete to increase resistance to freeze/thaw cycles. With use, however, other beneficial properties were noted, among which were:

1. Greater resistance to deicing chemicals
2. Improved workability
3. Lower water demand for a given slump
4. Allowable reduction in sand content

5. Reduced segregation and bleeding

6. Increased durability

7. Improved pumpability

Unlike entrapped air voids, which can absorb water, the billions of minute air bubbles per cubic yard of concrete *purposely* introduced into the concrete are not interconnected and are resistant to water penetration. These entrained air bubbles provide a reservoir which relieves the pressures of volume change inherent in concrete and prevents external and internal damage.

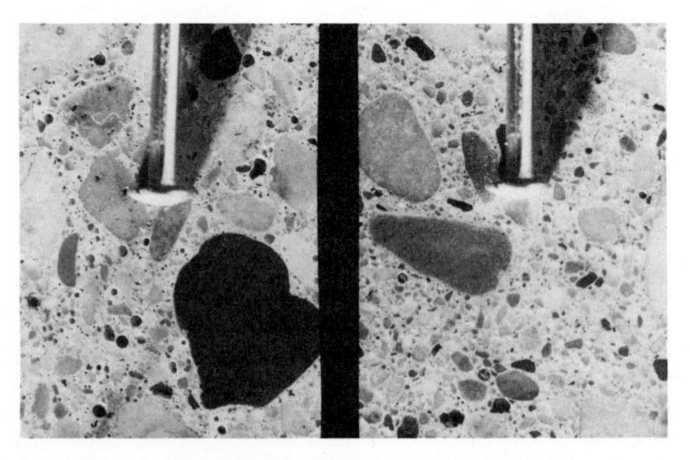

Figure 2-1 Polished section of air-entrained concrete as seen through a microscope. (Source: Portland Cement Association.)

The most frequently used type of air-entraining agent is a saponified Vinsol resin in an aqueous solution. Most manufacturers furnish their product in both a normal and a concentrated form, the concentrated form containing twice the amount of solids as the normal form.

The liquid dosages of air-entraining agents generally recommended per bag of cement are guides and are not to be considered exact under all conditions. The dosage may vary depending on the size and gradation of the aggregates, the richness of the mix, and the temperature of the concrete. Usually the normal dosage of 1 fl oz/100 lb of cement will result in the range of acceptable percentage. But only experience with specific, given materials will provide a reliable guide to the necessary dosage.

There is little question that air entrainment is big medicine, but, like any medicine, too much or too little will not produce the desired result. Moreover, once it is established, a desirable air content can be altered by other factors.

Low Air: Causes

Inadequate air entrainment, resulting from decreased air contents, may be caused by any of the following:

1. Excessive increase of the temperature of the concrete.
2. Prolonged mixing.
3. Insufficient mixing.
4. Incompatible admixtures. (In some cases the quality of one or both admixtures can be destroyed upon contact.)
5. Slump increase beyond 6-in. (after which air contents become unstable).
6. Very low slump concrete.
7. Faulty admixture dispenser.
8. Faulty air-test meter.

Low Air: Cures

1. Concrete temperature changes have a strong influence on entrained air percentages. When concrete temperatures are between 60 and 72°F the air content will vary slightly but will usually remain within an acceptable range. Temperatures above 75°F are almost certain to reduce air contents below requirements, and the higher the concrete temperature, the more drastic the reduction will be.

Taking the temperature of the concrete whenever an air test is being conducted will soon show a correlation between temperature and air content. Increasing the dosage of the air-entraining agent is necessary in higher temperatures, and if accurate records are kept of temperature, dosage, and resultant percentages, one can be guided as to what action to take on the reading of the concrete thermometer alone.

2. Prolonged mixing is frequently caused by mixing in transit during a long delivery run, by delays on the jobsite, or by the concrete's arriving on the jobsite earlier than requested, when the contractor is not ready to start. The prolonged mixing, increasing the concrete temperature by the heat of hydration and internal friction, will lower the slump and air content. A recent test at a jobsite indicated an air content of 6.1 percent air after 2 cu yd of concrete had been discharged. There was a delay of 1½ hours before the discharging could be continued. The air content of this same concrete had dropped to 2.7 percent.

Where long hauls are necessary, or slow pours are to be expected, the dry batch concrete supplier might well consider loading the aggregates and a quantity of water first, stopping the mixing drum, adding the cement by gravity, and mixing the load on the jobsite. This method may reduce the number of cubic yards normally carried, but it can pay off as a solution to the problem.

3. All too often, when the concrete delivery truck arrives on the job, the contractor insists on almost immediate discharge of the load. There are occasions when his anxiety, caused by a delay between delivery trucks and worry about a cold joint, is understandable. However, a few more minutes to mix the load properly will not usually affect the cold joint, while insufficient mixing may create more serious problems.

4. Not all admixtures are compatible. Where the different agents are introduced into the sand at the weigh scale, there is always the possibility of two coming into contact with one another to some degree. The concrete supplier can check this quickly by adding a small amount of one agent in a glass container and then adding the other. The result of the combination can be easily noted. Should there be a potential problem, the air-entraining agent may be added as described in item 6 following.

5. When slump is increased, the air percentage will increase. The maximum slump reached before the air decreases varies with the cement factor of the mix, the percent of sand to total aggregate, and the gradation of the materials. However, any slump over 6 in. is suspect.

6. It is difficult if not impossible to attain reasonable air percentages in very low slump concrete. The small air-entraining dosages require sufficient mixing water (slump) to dilute the air-entraining agent and disperse it evenly throughout the load in the field. Mixes designed in a laboratory are another matter. Here the air agent is introduced, in solution, into the mix to obtain the desired air content.

Most concrete suppliers add a percentage of the mixing water to the concrete truck along with the materials entering the mixer. This method aids in the loading process and saves valuable time on the jobsite, where additional water is added to obtain the required slump.

To reproduce, to some degree, the method used in the laboratory, introduce the air-entraining agent into the mixing water rather than into the fine aggregate, as is often done.

Dispense the air-entraining admixture into a graduated holding cylinder. Such cylinders are readily available from the admixture manufacturer. Install a normally closed valve at the bottom of the cylinder, and when the mixing water has been started into the truck, release the air admixture agent.

A ⅜-in. plastic line from the bottom of the holding cylinder leading down and ending at the discharge end of the water line will permit the air agent to mix with the water and enter the truck in solution. It is desirable to release the admixture as soon as possible after the water is started and to slow the discharge so that it is introduced into at least 80 percent of the water.

Generally the gravity flow of the admixture is too fast to accomplish the 80 percent figure. It may be necessary to reduce the ⅜-in. line with a foot of ¼-in. copper tubing at the discharge end. This copper tubing can easily be angled into the outflowing water. Further reduction of the flow can be created by pinching the end of the copper tubing to acquire the desired restriction.

7. There are several types of admixture dispensers on the market, from positive-displacement pumps to dispensers operated by compressed air. All the types are accurate and reasonably dependable. However, they must be properly maintained to remain accurate.

Dispenser output should be checked periodically and should certainly be checked immediately when low-air problems arise. Do not use water for this test, as the viscosity of the air-entraining agent is usually different. Use the admixture being dispensed for the test and keep a record of the output nearby.

The foot valve at the end of the intake line in the supply tank should be checked if low air is encountered. These valves are usually made of plastic or brass and are spring-loaded. Occasionally these springs lose their tension or are clogged with gook from the bottom of the supply tank and fail to close completely. Normally the liquid agent, after the pump stops, will drain back to the highest point in the discharge line. When the foot valve fails to close completely, the liquid can drain back to the level of the agent in the supply tank. Depending on the length of line from the highest point to the tank level, a loss of 20 percent or more can occur in the original output.

8. There is always the question of whether the air content is really low or not. When the concrete has the usual creamy look of the normally air-entrained concrete, it may well be that the air-testing procedure, or the air-testing equipment, is in error.

Poorly maintained dirty equipment should always be suspect. The pressure-type air meter is more accurate than the volumetric-type, but only when it is properly maintained. Using a small glass Chace meter that fits easily into the glove compartment of a car is simple and quick and provides a good check for air with average-slump concrete. It is certainly not unusual to have concrete rejected for low air when the fault was in the testing procedure.

Air-entraining admixtures are inexpensive, and that less of the fine aggregate is needed with their use more than compensates for the additional cost. In addition to the obvious improvement in concrete workability and reduction in bleeding that air-entraining admixtures bring about, tests have shown that air entrainment improves the concrete with aggregates of low durability.

Figure 2-2 shows the higher resistance of air-entrained concrete compared with non-air-entrained concrete when each has been exposed to freeze/thaw cycles. However, no picture of this kind can do justice to concrete under field conditions. The cement content, the aggregates, the slump, the percentage of air, and the severity of the freeze/thaw cycles vary greatly in the field. These field conditions plus heavy applications of deicing chemicals and heavy traffic can cause more rapid deterioration. Although controlled test specimens in a laboratory cannot duplicate field conditions exactly, they are a true indication of the value of air entrainment.

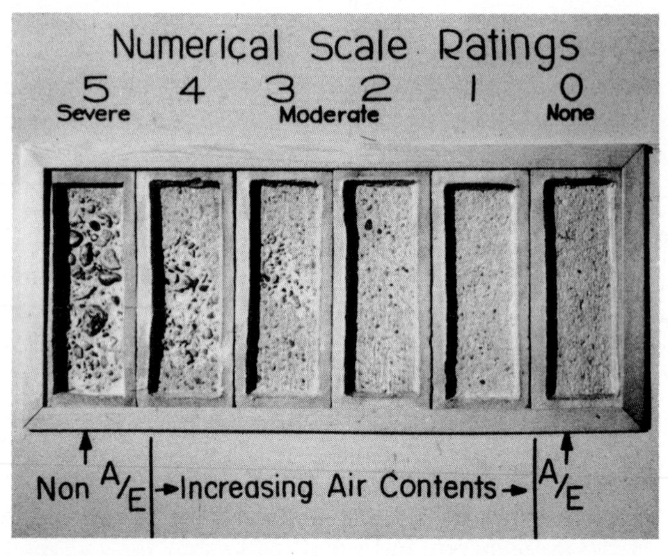

Figure 2-2 Test panels of concrete with no air-entrainment and with increased percentages of air under many cycles of freezing and thawing. (Source: Portland Cement Association.)

In areas where there are no, or infrequent, freeze/thaw cycles, concrete may be subjected to the attack of sulfate soil water or seawater. The severity of those attacks offers conclusive evidence of the advantages of using air-entrained concrete.

Many concrete producers include air-entraining admixtures in all their standard everyday concrete. However, the concrete supplier cannot be present on every job to check air contents on each load of delivered concrete. He is responsible only for air contents in the concrete as it leaves the truck. Changes that occur in the air percentage under placing practices, such as overvibration and faulty workmanship, cannot be considered the supplier's problem when cores taken later indicate lower air than requested. Under normal conditions the delivered air percentages will be within the acceptable range.

Concrete with normal air contents of 4 to 6 percent can usually be finished earlier than non-air-entrained concrete when the steel trowel is replaced with a magnesium trowel. However, when air contents are excessive, there is often a delay in the finishing operation to permit the excess air to escape from the surface. This phenomenon tends to occur more frequently when the concrete is placed on a cold subgrade.

High Air: Causes, Cures

Excessive air percentages present different problems than do low air percentages. High air content increases yield, reduces strength, and creates problems in the finishing operation, causing surface blisters and creating tackiness under the mason's knee boards.

Although some highway specifications permit air contents of 7 to 9 percent where freeze/thaw cycles are severe, or where heavy applications of deicing salts or chemicals are used, these air contents are too high for regular concrete. Highway concrete is placed differently from floor concrete. It is placed, screeded, smoothed and leveled with a bull float, and then burlap-dragged or broom-finished. Since no trowel is used in the final finishing operation, the problems of blistering and tackiness do not occur. These higher air contents do lower strengths somewhat, but the added durability on bridges and highways has proved the sacrifice of strength to be well worthwhile.

When excessive air percentages are encountered in standard building construction concrete, the cause of the problem must be found quickly.

Excessive air can result from several causes. The following possibilities should be considered:

1. Is the admixture dispenser working normally?
2. Has the air-entraining agent been changed to one that requires a lower dosage?
3. Could a delivery of air-entrained cement have been delivered in error? This is easily checked by half filling a small-diameter cylindrical jar, such as olives come in, with the cement in question, and adding water until the jar is two-thirds full. Cap the

jar and shake the contents vigorously, then place on a level surface and remove the cap. If the large bubbles that were created remain above the water for a great length of time, the cement is very likely air-entrained. With regular cement the bubbles dissipate almost immediately.

4. Has there been a change in the aggregate gradation? If daily gradation tests have been conducted, check for a serious deviation of the norm. Look for higher than usual percentages passing the No. 30 or No. 50 sieve.

5. Has the fine-aggregate wash water been considered? The presence of algae in water can, by itself, cause excessive air contents. When the fine-aggregate producer uses such polluted water in the sand production process, the algae entrapped by the sand will decompose, releasing a gas that will produce the high air phenomenon. Air tests conducted on concrete without an air-entraining agent, but made from such aggregates, recorded air (or gas) contents of over 12 percent.

Many aggregate producers wash their materials with water from ponds or small lakes. Hot, dry spells lower the water supply and create an atmosphere for abundant algae growth. The immediate solution for the concrete producer is to use sand from another source. The sand produced during the algae contamination should be set aside and turned over occasionally. Eventually it will be usable. Should the producer wish to continue production, he must install a suitable filter for the wash-water supply.

WATER-REDUCING ADMIXTURES

The amount of water added to concrete to make it placeable is far in excess of that required for the hydration of portland cement. Over 50 percent of this added water serves no other useful purpose than to lubricate the mix. This additional water adversely affects the shrinkage, durability, and strength of concrete. Water-reducing admixtures are added to concrete to reduce water demand for a given slump while retaining workability.

On the average, a reduction in total water content of 12 to 15 percent will attain this goal. Further, a mix with this reduced water content requires a lower cement factor to maintain the same W/C ratio.

The acceptance of Abrams's W/C ratio law led to the concept that density is a criterion of concrete quality created by the reduction of excess water. In general, density corresponds to unit weight. The heavier a cubic foot of a material is, the more dense that material is. When the Portland Cement Association investigated air-entrained concrete in the thirties, it found a decrease in the unit weight of such concrete. Further investigation showed that the introduction of air had reduced water demand and produced a more homogeneous, less permeable concrete, despite the lower unit weight.

The study further showed that an air-entrained mix would produce a lower 28-day strength than a non-air-entrained mix with the same cement factor. However, the desirable features of air could not be denied. It was later discovered that a water-reducing agent included with an air-entraining agent gave a much greater water reduction, improving strength and adding other desirable features in both the plastic and hardened concrete.

Benefits

Water-reducing admixtures (sometimes called *plasticizers*) offer the following benefits:

1. Reduced cost
2. Reduced bleeding, segregation, and honeycombing
3. Improved workability, particularly in wet harsh mixes
4. Reduced sand streaking in high-slump concrete
5. Increased strength
6. Reduced cracking and permeability
7. Reduced shrinkage
8. Increased bond of concrete to steel reinforcement

Chemical Properties

Water-reducing admixtures are made of (1) lignosulfonic acids and their salts, and/or modifications and derivatives of the same; or (2) hydroxylated carboxylic acids and their salts, and/or modifications and derivatives of the same; and (3) hydroxylated polymers.

Cement particles have a natural tendency to flocculate (clump together) and entrap water. Water-reducing agents act on the surface of cement particles, disperse the clumps, and release the previously entrapped water to the mix. Consequently the total water demand for a given slump is reduced.

History

In the late twenties and thirties, water-reducing agents were referred to as *cement-dispersing admixtures* and *water-reducing admixtures*. They were made of lignosulfonic acids obtained from the residue of wood-pulp plants. This raw material contained a small amount of glucose (sugar), and after processing, calcium chloride was added to the raw material to offset the retarding effect of the glucose. Dosages of the finished product were large in comparison to dosages of today's water-reducing agents. Usually 8 fl oz per sack of cement was recommended.

Rigid control was necessary in their use. The product was delivered dry, in bags. The precise number of gallons of water had to be added at the concrete plant and mixed to a consistent slurry. Careless mistakes were frequent, and improperly made solutions were the cause of many problems in the field.

Complaints of retardation, delaying the initial set for hours or more, were not uncommon. Nor was the fault always that of an improperly prepared solution. Sometimes the complaints stemmed from delivered material that had not been properly processed. Frequently the fault was not with the admixture at all but was created by improper field procedures in cool weather.

Even though the admixture did drastically reduce water demand and did permit a reduced cement factor, thus saving the concrete producer money, many users found the advantages were not worth the customer complaints.

It was a bad time for the water-reducing manufacturer. Although extensive research and development brought greater knowledge and better control over this product and made it reasonably trouble-free, it still carried the stigma of earlier problems. It was hard to sell.

Today, water-reducing admixtures are widely accepted, required in many specifications, and used in everyday concrete. Concrete producers, architects, and engineers consider them an important concrete ingredient. Reliable admixture manufacturers produce their products to conform to ASTM C 494, "Specifications for Admixtures for Concrete." It is the demand for conformity to this specification that assures the user that the admixture is acceptable.

Although millions of trouble-free yards of concrete are used annually with water-reducing agents, there are still occasions on which delayed sets occur.

Problems, Causes, Cures

Some of these complaints of delayed set can be attributed to the admixture itself, either by double-dosing or by an improper dosage due to a faulty dispenser. Excessive dosages of any admixture will result in concrete problems. Many retarded-set problems do not originate from the admixture but from certain temperature conditions under which the concrete is placed.

A bright sunny morning in the spring or fall does not always warm the subgrade that was subjected to freezing temperatures the night before. Fresh concrete at a temperature of 70°F can drop to below 50°F in a half hour or less when placed on such a cold subgrade. With or without an admixture, concrete will set more slowly when the temperature of the concrete falls below 50°F. If an admixture has been added to the problem concrete, there is a strong tendency to point the finger of suspicion in that direction, partially because these products had problems in the early stages of their development.

Externally heating the concrete, after placement, to induce quicker set in cold weather is of little help under the above conditions. This practice, if not properly done, can in fact create additional problems, setting the surface while the concrete below remains in the plastic state. The concrete may move under the power-troweling operation, causing an uneven surface finish. Frequently, when the concrete surface sets initially before the lower portion of the mass, *curling,* i.e., a lifting of the perimeter of the slab, is created. Warming the subgrade before placing the concrete eliminates the possibility of curling and can save hours of overtime on the finishing operation.

Some water-reducing admixtures contain air-entraining agents. Certainly the use of both these agents in concrete is a plus, but it has been found that adding them separately is preferable. The air content can then be raised or lowered without changing the dosage of the water-reducing agent. It is important not to exceed the manufacturer's recommended dosage of the water-reducing admixture.

Occasionally the recommended admixture dosage is exceeded. Even if the dispenser is accurate, there is still the possibility that the weigh batcher double-dosed the concrete batch. The batcher is human and may have inadvertently added the admixture twice. This is frequently the case when only one load in many fails to perform exactly as the rest. Further, the possibility that the cement weight was improperly batched should not be overlooked.

Creep, volume change, modulus of elasticity, durability, etc., are important qualities of concrete. It is generally assumed that where the strength of the concrete is adequate, these other quality features are present. Tests have shown that water-reducing admixtures along with an air-entraining agent add to these desirable features.

Super Plasticizers

Recently, new admixtures referred to as *super plasticizers* have been introduced to the concrete industry. These new admixtures greatly increase the water reduction and workability of concrete. The manufacturers recommend that these products be added to the concrete truck mixer at the jobsite, because of rapid slump loss.

When added to the mix these super plasticizers can produce high-slump concrete without risk of segregation or bleeding. The concrete, when placed, need not be vibrated. This one feature seems to be very desirable in situations in which one is placing concrete in narrow forms and in which reinforcing bars prevent good consolidation.

It is not fully known, at this writing, what other qualities these super plasticizers may impart to the concrete. The severe loss of slump may be a drawback if there should be a delay in placing the concrete after the admixture has been introduced into the mix. Recent tests of three of these new super plasticizers showed the concrete varied somewhat in compressive and flexural strengths, and in slump loss.

When perfected, these new products could very well be the water reducer of the future. At present they should be considered only where their high cost is justified. Until these new super plasticizers are better known, we must content ourselves with the more familiar water-reducing agents available.

Today's water-reducing agents are low in dosage at 2 to 3 oz/100 lb of cement. They are delivered in 55-gal drums or, for economy, in bulk of hundreds or thousands of gallons. Many admixture manufacturers will include bulk-holding tanks and dispensing equipment at little or no cost to the customer.

Most admixtures necessitate a change in mix design. The yield of a mix is decreased with a water reduction and must be compensated for in the new mix. Conversely, the introduction of an air-entraining agent increases yield and necessitates a different adjustment. Generally the manufacturer's representative is quite capable of readjusting original mix designs to change the mix to comply.

Here again, as with air agents, if there is not enough water in the concrete mix to dilute and disperse the water-reducing agent throughout the truckload of concrete, the value of the admixture is lost.

SET-RETARDING ADMIXTURES

Benefits

Set-retarding admixtures, although reducing water demand to some degree, are used primarily for delaying setting time in hot weather, in high temperature concrete, in large-mass concrete, or when pumping or hauling concrete long distances.

Dosages

The dosages, usually in liquid form, range from 2 to 5 oz/100 lb of cement, depending on ambient temperature and job conditions. The formulation of the admixture may vary slightly from one manufacturer to another but all should comply with ASTM C 494. The rate of delayed time of set is determined by the dosage. Usually delayed set lasts for only a few hours and permits normal form-removing procedures. Within 48 hours the concrete strength is equal to or higher than that of plain concrete.

When concrete loses its plasticity, it is considered to have reached the vibration limit. When concrete temperatures or ambient temperatures are high, set retarders increase the normal allowable time to reach the vibration limit. This time delay is often helpful in eliminating the possibility of a cold joint. Retarders are useful when placing concrete in thin sections, as in shells and on bridge decks, where full dead load is achieved before the concrete makes its initial set.

Retarding agents, often referred to as *water-reducing set-retarding admixtures,* decrease the natural interparticle attraction between cement grains and release this normally entrapped water to become a part of the mixing water.

It is important that the recommended dosages for set-retarders be followed closely; moreover, safeguards against possible double dosages should be provided. In one case, someone added 1 gal (128 oz) of retarder to 1 cu yd of concrete, mistaking it for calcium chloride. The concrete remained plastic for 5 days. The 28-day strength test of this concrete was over 7000 psi. Obviously, overdosing, in spite of excellent strength gains, must be carefully guarded against.

On hot, dry, windy days, concrete will set more quickly than under normal conditions. When these conditions prevail, some contractors will request the addition of a set retarder to avoid premature set. This permits the finishing operation on a large floor pour to proceed at a more leisurely pace and produces a higher quality floor than would have been obtained under the circumstances.

Problems

Set retarders serve a useful purpose, but they are not the solution to poor mix design, inferior materials, or low cement factors. They do act differently with some types and brands of cements, and caution is required when they are used. They should never be used in low slump concrete. Moreover, care should be taken that they are uniformly dispersed throughout the mix. If they are not, concentrations of the admixture will remain in pockets and create areas that can remain plastic long after other sections of the pour have set.

ACCELERATING ADMIXTURES

Accelerating admixtures are used in concrete to shorten the time of set and to increase early strength development. Several chemicals will accelerate the hardening of the cement paste, calcium chloride being the most popular one in common use.

Calcium Chloride

Calcium chloride is sold in flake or granular form and should always be introduced into concrete in solution. The instructions for making the solution, stated on each bag, should be followed carefully. If instructions are not readily available, a solution may be prepared by adding 4 lb of calcium chloride to 1 gal of water.

Calcium chloride is most generally used in cold weather to hasten setting time and permit earlier finishing. Users of this chemical should not, as is often the case, consider it as an antifreeze, since concrete placed under cold weather conditions *can freeze,* even with calcium chloride added.

Problems

Further, using calcium chloride can affect other characteristics of the concrete.

1. It generally increases the drying shrinkage.
2. It can lower the resistance of the concrete to freeze/thaw cycles and to the attack of sulfates.
3. It induces temperature rise and can increase internal stresses.
4. It attacks aluminum conduit and can promote corrosion in reinforcing steel where adequate protection of the steel is not provided.

Because of these negative features, many concrete design engineers will not permit the use of calcium chloride in their specifications. Most state and federal agencies prohibit its use. However, when a building contractor chooses to continue construction under cold weather conditions, many agencies will accept the addition of 100 lb of cement per cubic yard of concrete as a safe substitute for the calcium chloride.

If calcium chloride is to be used, a maximum of 2 percent by weight of the cement is recommended, with 1 or 1½ percent being preferable. When calcium chloride is used, the normal protective procedures for cold weather concrete should be followed.

Frozen Concrete

Frozen concrete is a serious problem that is often more obvious with the spring thaw. Water, expanding some 9 percent when frozen, destroys the concrete from within, causing it to disintegrate and to spall beyond repair.

To prevent concrete from freezing in doubtful weather, refrain from using high-slump concrete. When the surface of the placed concrete can be walked on, cover the concrete with at least 4 in. of salt hay. Be sure to spread the hay at least 6 in. beyond the perimeter of the pour and cover it all with polyethylene sheeting.

If the concrete does freeze, cover with a tarp and apply steam. If the frost injury is noticed soon enough, this procedure often saves the concrete.

POZZOLANS: COMPOSITION

A *pozzolan* is a siliceous material that reacts with the liberated lime of the cement in the presence of moisture to form new cementitious compounds. It is generally used as a substitution for a percentage of portland cement to obtain certain desirable features. Pozzolans are helpful where the concrete aggregate is alkali-reactive.

Although concrete containing pozzolans gains strength more slowly than straight portland cement concrete, its ultimate strength is equal to, or exceeds, the strength of the latter. When there is a cement shortage, pozzolans, because they are less costly than cement, are looked upon with more favor than they were in past years.

Fly Ash: Composition

Power plant ash, known as *fly ash,* is the most popular of these materials. A finely powdered residue from burning coal, it is composed largely of compounds of silica, alumina, iron, and lime. The quality of fly ash is dependent on the source of the coal and on the type of boiler from which the ash is produced.

Concrete made with the addition of high-quality fly ash shows no evidence of inferior concrete quality. The control of the carbon content in fly ash production is important since high carbon contents can drastically reduce or eliminate air entrainment. One should be sure the fly ash producer being considered has a record of quality performance.

Although the cost of fly ash is low compared to that of cement, there are initial startup expenses. A separate silo may have to be erected, and possibly a new weigh scale and hopper will be necessary. If these cannot be conveniently added to the present plant, a second shop setup may be required.

These initial setup expenses will pay for themselves in time. If you are considering the use of fly ash, speak first to the concrete supplier who has been using it for some time. In your area it may or may not be worthwhile.

A composite report of several recent studies on fly ash indicates that the fineness, chemical composition, and physical properties of fly ash vary depending on the source of the coal, the burning method, and the combustion equipment. As most plants vary in their source of supply and methods of manufacture, there naturally will be variations in the finished product.

Used as a cement replacement, fly ash can impart desirable features to the concrete. In hot weather or mass concrete, fly ash will reduce the heat of hydration. In many cases the strength of the concrete is improved.

Comments

Although the total water content of the original mix will remain fairly constant, a redesign of the mix will be necessary when fly ash is used as a cement replacement. Adjustment in the air-entraining admixture may be necessary. Be sure to adjust the yield of the mix because of the difference in specific gravity of the fly ash.

3 / CONCRETE MIX DESIGN

It is the intent of this chapter to explain to those unfamiliar with the design of concrete mixes an easily understood procedure for proportioning materials for a concrete mix. Concrete is simply a combination of properly proportioned fine and coarse aggregate held in suspension by the cement and water paste. The proper proportion and quality of these materials will produce the desirable characteristics of the plastic and hardened concrete.

DESIGN DEFINITIONS

Listed below are some fundamental facts and definitions that should be reviewed before proceeding to a consideration of mix design:

Unit weight = weight of 1 cu ft of a material

Unit weight cement = 94 lb

Unit weight water = 62.4 lb

One gal water = 8.33 lb

One cu ft water = 62.4/8.33 = 7.5 gal

One gallon of water increases slump approximately 1 in.

A 1-in. increase in slump creates a loss of approximately 200 psi.

sp gr — specific gravity; the ratio of the weight of a material to the weight of an equal volume of water. (See ASTM C 127, "Test for Specific Gravity and Absorption of Coarse Aggregate.")

S/A ratio — sand/aggregate ratio; the percentage of sand to total aggregate.

SSD	saturated surface dry; no absorption.
cu ft	cubic foot.
cu yd	cubic yard; 1 cu yd = 27 cu ft.
W/C ratio	water/cement ratio; number of gallons (or pounds) of water used per bag of cement.

ABSOLUTE VOLUME

Absolute volume refers to the volume without voids. If we were to melt 1 cu ft of cement, the resulting mass would theoretically be without voids. As we are unable to do this, we are given a mathematical formula for accomplishing the purpose.

EXAMPLE 3-1

$$\frac{\text{Wt of 1 cu ft of material}}{\text{Sp gr of material} \times \text{unit wt of water}} = \text{absolute volume}$$

$$\frac{\text{94 lb cement}}{3.15 \times 62.4 \text{ lb}} = 0.48 \text{ (52\% voids)}$$

For simplicity of explanation let us assume we have a 1 cu ft container filled with coarse aggregate. Here we have loose or dry-rodded volume with voids. We will now add SSD sand to eliminate these large voids, leaving instead a far greater number of smaller voids. These in turn are then filled with a water and cement paste to eliminate the sand voids. The resultant mass would be absolute volume.

It will be found that the larger the maximum size of coarse aggregate the less sand is needed to fill the voids. The S/A ratio will very depending on the gradation variation of the same-size aggregate. Increasing the sand increases the surface area, demanding more water to achieve a required slump and necessitating more cement to meet desired strength.

In the early days of jobsite concrete, the concrete was mixed by loose volume rather then by weight. A 1-2-4 mix, i.e., a 1-6 mix (1 part cement to 6 parts aggregate), consisted of 1 shovel of cement, 2 of sand, and 4 of coarse aggregate. Water was added to create the desired slump. For a stronger concrete a 1-2-3 mix would be used (1 part cement to 5 parts aggregate).

ABRAMS'S LAW: WATER/CEMENT RATIO

No one could be certain what strength would be attained with the loose volume method. Water was added by hose and slumps varied. After much testing, a Mr. Duff Abrams proposed the relationship of water to cement, and the W/C ratio law is used to this day. The table in Example 3-2 takes into account the different aggregates, and the different sand/aggregate ratios used.

EXAMPLE 3-2

Before continuing we should clarify the idea of W/C ratio by weight. In many other countries, and increasingly too in the United States, the expression "gallons per sack" (gps) is not the criterion for mix design. The W/C ratio *by weight* is the standard (W/C refers to pounds of water divided by pounds of cement).

W/C by wt.	*Gal per sack*	*Approximate 28-day strength, psi*
.45	5.0	5000
.49	5.5	4500
.53	6.0	4000
.57	6.5	3500
.62	7.0	3000
.66	7.5	2500
.71	8.0	2000

To convert the W/C ratio by weight to gallons per sack, terms we might be more familiar with, we use the following formula:

$$W/C \times \text{lb/sack} = \text{lb water} \div \text{lb/gal} = \text{gps}$$

To find the gallons per sack to make a mix with a W/C ratio of .62, we apply the formula:

$$.62 \times 94\ \text{lb/sack} = 58\ \text{lb} \qquad \frac{58\ \text{lb}}{8.33\ \text{lb/gal}} = 7.0\ \text{gps}$$

Conversely,

$$\frac{W}{C} = \frac{7.0\ \text{gps} \times 8.33\ \text{lb/gal}}{94\ \text{lb/sack}} = \frac{58\ \text{lb}}{94\ \text{lb/sack}} = .62$$

For the sake of simplicity the W/C ratio will be expressed in gallons per sack in the following pages. It should be noted that the anticipated 28-day strength, for a given number of gallons per sack, is approximate. Because strength depends in large part on the type of aggregate and the various S/A ratios used, the strength of the mix can not be predicted accurately.

DESIGN OF A 1-2-4 MIX

We will design a 1-2-4 mix, which was generally assumed to produce a 2500-psi mix. Using Example 3-1 as a guide:

$$\frac{\text{Wt of 1 cu ft of material}}{\text{Sp gr of material} \times \text{unit wt of water}} = \text{absolute volume}$$

Ingredient		*Volume, cu ft*		*No. parts*		*Yield, cu ft*
Cement	$\frac{94 \text{ lb}}{3.15 \times 62.4 \text{ lb}}$	= .48	×	1	=	.48
Sand	$\frac{95 \text{ lb}}{2.65 \times 62.4 \text{ lb}}$	= .574	×	2	=	1.15
Gravel	$\frac{105 \text{ lb}}{2.65 \times 62.4 \text{ lb}}$	= .635	×	4	=	2.54
Water	$\frac{7.5 \text{ gal}}{7.5 \text{ gal/cu ft}}$				=	1.00
						5.17

$$\text{Cement factor} = \frac{27.00 \text{ cu ft/cu yd}}{5.17 \text{ cu ft/sack}} = 5.2 \text{ sacks cement}$$

Given the aggregates used in the 1-2-4 mix described, it was found that the cubic yard of concrete required 5.2 bags. We now know the cement factor and the number of gallons needed per cubic yard, that is, 5.2 sacks × 7.5 gps = 39 gal.

Cement $5.2 \times .48 = 2.49$ cu ft
Water $39 \div 7.5 = 5.20$ cu ft
7.69 cu ft yield (cement + water)

Specific gravity of both aggregates = 2.65

27.00 cu ft
−7.69 cu ft (cement + water)
19.31 cu ft × (2.65 × 62.4) = 3186 lb total aggregate ÷ 6 parts
= 531 lb

Sand	531 × 2 =	1062 lb
Gravel	531 × 4 =	2124 lb
		3186 lb total aggregate for 1 cu yd

Thus, for 1 cu yd of 1-2-4 concrete, we would need:

Cement (SSD, lb)	489 (5.2 bags × 94 lb/bag)
Sand (SSD, lb)	1062
Gravel (SSD, lb)	2124
Water (gal)	39

and our proportions check:

$$\frac{489}{489} : \underline{1.0} \qquad \frac{1062}{489} : \underline{2.17} \qquad \frac{2124}{489} : \underline{4.34}$$

We have now designed a yard of 1-2-4 concrete. The S/A ratio (1062 ÷ 3186) is 33.3 percent, and the final weights are reasonably close to our planned 1-2-4 mix. If the coarse aggregate were 1½ in., this mix might be fairly acceptable. If the ready-mix supplier uses ¾-in. aggregate, the mix might be too harsh and unacceptable.

Although the mix we designed, with well-graded aggregates, would give us the maximum strength with the required cement factor, we would usually modify it. An adjustment in the S/A ratio, increasing the sand content, would add to the workability and be more acceptable to the everyday customer.

The sand content of total aggregate as we designed it is 33.3 percent. For the second trial mix we could easily increase the S/A ratio to 36.0 percent as shown in Examples 3-3*a* and 3-3*b*. The adjustment in Example 3-3*a* is by weight while that in Example 3-3*b* is by volume.

EXAMPLE 3-3*a*

Sand	3186 lb × 36 percent	= 1147 lb
Gravel	3186 lb − 1147 lb	= 2039 lb
		3186 lb total aggregate

EXAMPLE 3-3*b*

Sand 19.31 cu ft × 36 percent
= 6.95 cu ft × 2.65 × 62.4 = 1147 lb

Gravel 19.31 cu ft − 6.95 cu ft
= 12.36 cu ft × 2.65 × 62.4 = 2039 lb

When both fine and coarse aggregates have the same specific gravity, the total weights are constant. Should the specific gravity of the coarse aggregate be different from that of the fine aggregate, the weights must be based on the volume method.

Let us assume we are using crushed stone with a specific gravity of 2.82 for the coarse aggregate. Keeping the cement and water constant, for simplicity, we would determine the weight of the coarse aggregate as in Example 3-3*c*.

EXAMPLE 3-3*c*

Sand	19.31 cu ft × 36 percent =	6.95 × 2.65 × 62.4 = 1147 lb
Stone	19.31 cu ft − 6.95 cu ft =	12.36 × 2.82 × 62.4 = 2175 lb

It should be noted when comparing Example 3-3*b* with Example 3-3*c* that the sand weight remained constant while the coarse aggregate weight changed. With a specific gravity of 2.65 in Example 3-3*b*, the coarse aggregate weight was 2039 lb; with the specific gravity of the stone at 2.82 in Example 3-3*c*, 2175 lb was necessary to keep the yield constant.

In actual practice, of course, the sand would have to be increased with the use of crushed stone for workability because the angular structure of crushed aggregate reduces the flowability of a mix. Usually an increase of 3 percent sand, by volume, will be required when switching from gravel to stone, with a similar volume percentage reduction from the stone.

DESIGN OF AIR-ENTRAINED CONCRETE

In the early days it was not necessary, as it is now, to figure the effect of air entrainment in a mix design. In Example 3-3*d*, we have again kept a constant cement and water factor but allowed for the inclusion of 5.0 percent of air in the design.

EXAMPLE 3-3*d*

Cement	5.2 sacks × .48 cu ft =	2.49 cu ft
Water	39 gal ÷ 7.5 gal/cu ft =	5.20 cu ft
Air	5% × 27 cu ft =	1.35 cu ft
		9.04 cu ft

27.00 cu ft − 9.04 cu ft = 17.96 cu ft total aggregate

To continue our example, we will use sand with a specific gravity of 2.65 and gravel with the same specific gravity.

Sand	17.96 cu ft × 33.3 percent	= 5.98 cu ft
Gravel	17.96 cu ft − 5.98 cu ft	= 11.98 cu ft

Sand	5.98 cu ft × 2.65 × 62.4 =	987 lb
Gravel	11.98 cu ft × 2.65 × 62.4 =	1977 lb

If the coarse aggregate were crushed stone with a specific gravity of 2.82, we would have to increase the sand content by 3.0%, from 33.3% to 36.3%, for workability.

Sand	17.96 cu ft × 36.3 percent =	6.52 cu ft
Stone	17.96 cu ft − 6.52 cu ft =	11.44 cu ft
Sand	6.52 cu ft × 2.65 × 62.4 =	1076 lb
Stone	11.44 cu ft × 2.82 × 62.4 =	2013 lb

The above exercises were given merely to show the theory of designing air-entrained concrete mixes. They should be considered as *trial* mixes. In actuality, adjustments to these figures would be necessary. Concrete mixes with air require a lower percentage of sand than mixes with no air. The billions of air bubbles act, in effect, as fine aggregate in the mix, and with a reduced sand content, will retain or improve the same workability of the no-air mix.

Normally, the introduction of air entrainment to a mix will reduce the 28-day compressive strength unless the cement factor is increased to compensate. However, this is not always true in the case of very lean mixes, which will occasionally show an improvement in strength. This phenomenon is related to the much lower water demand of the air mix to equal the slump of the no air mix.

There is a general rule of thumb for finding how much cement to add to a mix for each percentage of increased air over the no-air mix to maintain equal strength. No concrete, however, is without some air. There is always ½ to 1½ percent of entrapped air in plastic concrete, so that the total air content of an air-entrained mix is not entrained air alone. Frequently the cement content of the no-air mix is so much higher than necessary to meet strength requirements that the lower-strength air mix needs no adjustment in cement factor.

Should it be necessary to compensate for the reduction in strength with the introduction of air entrainment, adding a water-reducing agent to the mix, rather than increasing the cement factor, should be considered. A water-reducing admixture will lower the water demand between 10 and 15 percent for a given slump. This reduction of the W/C ratio may increase the strength, so much so that a *reduction* rather than an increase in the cement factor may be possible. Besides the obvious reduced cost factor, a concrete mix containing both an air-entraining and a water-reducing agent provides features to the plastic and hardened concrete not otherwise attainable.

High compressive strength (high cement content) does not necessarily guarantee high quality. Durability and workability must be considered. Higher cement contents than necessary, particularly in warm weather, can create problems that do not develop in leaner mixes. The introduction of a water-reducing agent, if permitted, can reduce a high cement content in concrete to a more reasonable figure and eliminate some of the warm weather problems.

In concrete of low slump ranges, the air-entraining agent and the water-reducing admixture will not perform as well in the field as they did in the testing laboratory. In the laboratory these agents are usually introduced into the mix as part of the mixing water. In the field, if there isn't enough water in the mix (low slump) to dilute these agents and disperse them throughout the concrete truckload, the air will fall short of required specifications while the water reduction will fall short of expectations.

Unless otherwise specified, the normal percentage of air desired in a concrete mix is between 4 and 6 percent. Keep cement factors to the minimum that meets required strengths, with a reasonable safety factor. When adding an air-entraining agent to a no-air mix, a good rule of thumb to follow is to reduce ¼ gal of water and 5 lb of sand per sack of cement from the mix for each 1 percent of entrained air. In Example 3-3*e*, we will redesign a no-air mix using this rule of thumb.

REDESIGN OF NO-AIR MIX FOR AIR ENTRAINMENT

EXAMPLE 3-3*e*

Using a 5.0 sack mix and designed for 2500-psi concrete, the *no-air* mix was as follows:

Ingredient	*Amount*	*Yield, cu ft*
Cement (SSD, lb)	470	2.40
Sand (SSD, lb)	1296	7.84
¾-in. Gravel (SSD, lb)	1945	11.76
Water (gal)	37.5	5.00
		27.00 (=1 cu yd)

The no-air mix just listed, as redesigned for the addition of an air-entraining admixture figuring 5 percent would be as follows:

Ingredient	*Amount*	
Cement (SSD, lb)	470	(unchanged)
Sand (SSD, lb)	1271	(decreased from 1296)
Water (gal)	36.25	(decreased from 37.5)

The yield would be figured as follows:

Ingredient	*Amount*	*Calculation*	*Yield, cu ft*
Cement (SSD, lb)	470	÷(3.15 × 62.4)	2.40
Sand (SSD, lb)	1271	÷(2.65 × 62.4)	7.68
Water (gal)	36.25	÷7.5	4.83
Air	5.0%	×27	1.35
			16.26
Gravel (27 cu ft − 16.26 cu ft)			10.74
			27.00 (=1 cu yd)

(In the calculations, 3.15 is the specific gravity of cement, 62.4 lb is the weight of 1 cu ft of water, 2.65 is the specific gravity of sand and of gravel, 7.5 is the number of gallons per cubic foot of water, and 27 is the number of cubic feet per cubic yard.)

The new mix with air is as follows:

Ingredient	*Amount*	*Yield, cu ft*
Cement (SSD, lb)	470	2.40
Sand (SSD, lb)	1271	7.68
¾-in. Gravel (SSD, lb)	1776*	10.74
Water (gal)	36.25	4.83
Air	5.0%	1.35

* SSD weight of gravel is calculated thus: 10.74 cu ft × (2.65 × 62.4) = 1776

This new mix design would be referred to as a *trial mix* and should not be used in construction until it has been studied in the plastic state for workability, slump, and air content. Minor adjustments should be made, if necessary, and then a second trial mix should be made so that cylinders can be cast to test for strength results.

DESIGN OF TRIAL BATCHES

When these mixes are designed in a testing laboratory, they are usually made up in *1-cu-ft* batches, mixed in a portable mixer, studied, and tested. Designed at a concrete plant, the first trial mix is of a *1-cu-yd* batch in a concrete truck mixer. However, when concrete cylinders are to be cast, the truck mixer should never have less than *2* cu yd for a test batch.

When the trial batches are fully acceptable, we can continue to design the 2000-, 3000-, and 3500-psi mixes using the new 2500-psi mix as a guide. Half a sack of cement (47 lb) per cubic yard is generally considered sufficient to increase the compressive strength of a mix by 500 psi.

In Example 3-3*f*, we will keep the coarse aggregate constant. The only change will be in the content of the cement and in the fine aggregate. When the cement is reduced, the sand is increased; when the cement is increased, the sand is reduced. This is an acceptable procedure as the cement fines in a mix act as fine aggregate.

EXAMPLE 3-3*f*

Ingredient	*2000 psi*	*2500 psi*	*3000 psi*	*3500 psi*
Cement (SSD, lb)	423	470	517	564
Sand (SSD, lb)	1310	1270	1230	1190
¾-in. Gravel (SSD, lb)	1776	1776	1776	1776
Water (gal)	36.25	36.25	36.25	36.25
Air (%)	5	5	5	5
Slump (in.)	5	5	5	5
W/C (gps)	8.05	7.25	6.59	6.04
W/C (by wt)	.71	.64	.58	.54

$94 \div (3.15 \times 62.4) = .48$ cu ft ½ bag $= .24$ cu ft

$.24 \times (2.65 \times 62.4) = 40$ lb sand

It will be noted in the mix designs in Example 3-3*f* that the sand content was *decreased* 40 lb as the cement factor rose above the 2500-psi guide mix and was *increased* 40 lb in the 2000-psi mix, because of the lower cement factor. This 40 lb of sand was computed as follows to maintain yield:

Cement sack $= 94$ lb $\div (3.15 \times 62.4) = .48$ cu ft (½ sack $= .24$)

$.24 \times (2.65 \times 62.4) = 40$ lb sand

In actual practice the air contents and the total water to retain a 5-in. slump will vary somewhat. Theoretically, if the 2500-psi mix met requirements, the other mixes should be in the ball park. Actually, no mix, no matter how well designed, will perform exactly the same with any consistency. Temperature, job conditions, a change of cement brand, etc., will affect jobsite performance. Still, a well-designed mix will perform far more consistently than a poorly designed mix.

DESIGN OF AIR-ENTRAINED CONCRETE INCLUDING A WATER-REDUCING ADMIXTURE

If we wish to introduce a water-reducing agent to any of the mixes designed in Example 3-3*f*, the exercise in Example 3-3*g* should be followed. Water-reducing agents vary between 10 and 15 percent. Using an average of 12 percent as an assumed average, we will redesign as shown in Example 3-3*g*.

EXAMPLE 3-3*g*

We start with the 2500-psi air-entrained mix from Example 3-3*f*:

Ingredient	*Amount*	*Yield, cu ft*
Cement (SSD, lb)	470	2.40
Sand (SSD, lb)	1270	7.68
¾-in. Gravel (SSD, lb)	1776	10.74
Water (gal)	36.25	4.83
Air (%)	5.0	1.35
		27.00 (= 1 cu yd)

Leaving the cement and sand factor constant, we reduce the water by 12 percent (4.32 gal) and we now have 36.25 − 4.32 = 31.93, or 32.0 gal.

Ingredient	*Amount*	*Yield, cu ft*
Cement (SSD, lb)	470	2.40
Sand (SSD, lb)	1270	7.68
Water (gal)	32.0	4.27
Air (%)	5.0	1.35

27.00 cu ft − 15.70 cu ft = 11.30 cu ft gravel
11.30 cu ft × (2.65 × 62.4) = 1868 lb gravel

It will be noted that the coarse aggregate weight increased 92 lb over the weight of the original air-entrained mix. This was necessary to compensate for the water reduction and to maintain yield. At this point we question the value of the use of a water-reducing agent as it increases the cost of the mix, plus the cost of the additional gravel.

However, a further study of the new mix will reveal a decided *plus* for the addition of the water-reducing agent. For the *new* mix we have:

Ingredient	*Amount*	*Yield, cu ft*
Cement (SSD, lb)	470	2.40
Sand (SSD, lb)	1270	7.68
¾-in. Gravel (SSD, lb)	1868	11.30
Water (gal)	32.0	4.27
Air (%)	5.0	1.35
		27.00 (= 1 cu yd)

We then calculate the W/C ratio by the methods given in Example 3-2:

W/C ratio	*New mix*	*Air mix*
In gps	6.4	7.25
By wt	.56	.64

The addition of the water-reducer has decreased the W/C ratio to the point where it is similar, theoretically, to that of a 3000-psi mix. With the cost of cement being far higher than the combined prices of both the increase in gravel and the water-reducing agent, a small profit can be realized.

The new mix, with the lower cement factor and water-reducing agent, will in no way reduce the quality of the original air-entrained mix. It can, in fact, improve the quality of the original mix of equal slump. Further, it may be expected to increase the 28-day compressive strength. An air-entrained mix with a water-reducing agent is most helpful when the concrete mix is to be pumped.

METHODS OF ORDERING CONCRETE

In some areas concrete is *not* ordered by strength but by volume (1-2-4, 1-2-3, etc.), and in others by bag content (5 bags, 5½ bags, 6 bags, etc.).

The customer in such cases really has no complaint if the concrete ordered in these terms does not come up to the strength he anticipated. In fact, the 1-2-4 mix ordered is not a truly proportioned 1-2-4 mix but a modified one to make the mix workable. A truly proportioned 1-2-4 mix, by weight, would usually result in a harsh, bony, unworkable mix.

In the case of the customer who orders by bag content, the concrete producer must provide the necessary cement factor. Because of the additional cost, this would normally prevent the producer from adding a water-reducing agent unless he could reduce the cement factor to compensate. But there are some customers, although their number has greatly decreased in the past few years, who would prefer that *no* water-reducing agent be included in their concrete. This is a holdover from the earlier days when the water reducers were erratic. Today, not only customers but also architects and engineers specify the inclusion of water-reducing agents.

A word of caution. It cannot be assumed that, because one knows the mathematical formula for designing concrete mixes, one can therefore guarantee the compressive strength of that design mix. The aggregate size, particle shape, soundness, and gradation play an important role in the quality and durability of the concrete in the plastic and hardened state.

One concrete supplier, with a 5-bag mix at a 4-in. slump, may attain 3000 psi or higher, while another with poorer materials can be lucky to achieve 2500 psi. Even with the same aggregates, the cement brand plays an important part in ultimate strength. Raw materials used in the manufacturing process of cement vary because of different plant locations. It is only natural that this variation in raw materials will influence the strength of the concrete to some degree.

A general rule of thumb is to expect the 7-day cylinder result to be approximately 60 percent of the anticipated 28-day result. This is not true in all cases, as some cements may produce much lower 7-day strength results and yet be well over the expected 28-day results. Conversely, some cements with higher 7-day results than normal may gain little additional strength in the following 21 days.

HIGH CEMENT FACTOR

It is usually assumed that the higher the cement factor, the higher the strength and quality of the concrete. Generally, this is true. But cement factors far in excess of those needed to obtain a given strength are costly and are frequently the cause of unnecessary problems, particularly in hot weather. Of course, the W/C ratio on specification work must be followed, but standard everyday concrete should be designed with only a reasonable safety factor.

Most major project specifications require laboratory test results to be 15 percent higher than required strength. This is desirable from the point of view that field conditions can never be as ideal as controlled conditions in a laboratory. A few design engineers require a 25 percent safety factor, while others may specify 3000-psi concrete yet specify a W/C ratio that may produce 4500 psi.

However well-intentioned this much higher than necessary cement factor is, it can well cause structural defects. Concrete shrinkage cracks are caused by the volume change in the cement paste, not in sound aggregates. An excessively rich paste may well develop more shrinkage cracks and bigger volume changes at later dates than concrete with more reasonable cement factors.

The richer the mix, the higher the heat of hydration. High concrete temperatures can result in rapid slump loss, in reduced air percentages, and frequently in partial hydration and in a reduction in proper curing, all of which decrease the quality of the hardened concrete.

It is generally assumed that where strength is present, other quality features are also present. A project file, full of high-strength concrete results, may be satisfying, but strength alone is no guarantee of concrete durability.

4 / CONCRETE STRENGTH

COMPRESSIVE STRENGTH: HOW TO DETERMINE

The compressive strength of concrete is determined by molded concrete cylinders (6 in. × 12 in.) carefully made, sampled from the ready-mix truck at the point of discharge, placed and protected on the jobsite for 24 hours in a prescribed manner. After 1 day in the field they should be *carefully* transported to a testing laboratory, stripped and placed in a curing room at controlled temperatures, and capped and tested at the required age at a given loading rate.

If any one step in the above procedure is not conducted by the standard method, the test can reflect lower strengths than the true strength of the concrete as delivered. Although this is one of the most important field tests on a concrete job, the prescribed methods are frequently not followed and problems result.

There are, of course, other factors that influence the ultimate strength of concrete. It is the intent of this chapter to examine the reasons for concrete strength variations, particularly low strength results.

A customer ordering concrete should always specify the strength required, and will have little basis for complaining when ordering a 1-2-4 mix or a 5-bag mix if that concrete fails to meet the expected strength. Even the customer who orders by strength *must* conform to certain good construction practices if he expects to obtain the strength requested. Most concrete producers guarantee their strength concrete only when the slump does not exceed 4 or 4½ in.

QUALITY CONTROL INSPECTION

Where there are concrete specifications, they call for (or specify) a minimum cement factor, a maximum-size coarse aggregate, a maximum slump, a certain percentage of air, and a particular type and quantity of any other admixture. If the building project is a large one, the mix designs and W/C ratio curves, from a testing laboratory, are included in the specifications. Further, quality control inspection, which would involve an inspector at the concrete plant and one in the field, may be mandated.

The plant inspector checks the accuracy with which the concrete is batched and conducts gradation tests of the aggregates for conformity to the specifications. The field inspector who takes slump tests and air tests, and makes concrete cylinders generally has the authority to reject loads with higher slumps or lower air contents than specified. Both are required to submit written reports of their daily tests to their superior, who forwards them to the client.

LOW STRENGTH PROBLEMS: CAUSES

On some jobs the general contractor may perform his own field tests. Occasionally he might assign one of his own personnel to take a slump test or to mold cylinders. Unfortunately, such personnel are not always qualified, and so it is not unusual to have a laborer with little experience rodding a cylinder with a ¾-in. rebar instead of with the ⅝-in. bullet-nosed steel rod required. There is a good chance that the cylinders molded in this manner will produce lower strength results than cylinders molded with the proper tool.

The concrete supplier now has to defend his product having little or no knowledge of what happened when the concrete was delivered a month before. Back in the early fifties, a list was published of over 50 different contributing causes of why some cylinders fail. Because of this array of possible causes, it is understandable if the concrete producer cannot readily supply a suitable answer.

Often, a study of the cylinder test report from the testing laboratory may provide a clue to the strength failure. The test report, along with the results of the compressive test, usually provides other pertinent data. It will always provide the date the cylinder was cast and the date it was tested. Was it a 7-day test or a 28-day test result? It is generally assumed that the 7-day strength result will be in the neighborhood of 60 percent of the anticipated 28-day result. However, some cements react differently from others, and a lower-than-expected 7-day test result may improve at a later age and more than meet the required strength at 28 days.

How many cylinders reported were cast on the same date? It should be recognized that it is not necessary that 100 percent of all tests exceed the minimum-strength requirement. Did all the cylinders of a set fail, or did just one? If, for example, only one cylinder failed out of a group of three, it could well be ignored. If all the cylinders in a single group failed, one should look further on the test report. Was the slump excessive? Were the cylinders cured in the field for several days before being transported to the testing laboratory? Were the cylinders cast on a Friday and left on the jobsite until Monday, or later? Were the weights on the cylinders, generally given on the test report, lower than usual? This could well be caused by high air contents. What was the reported air content?

What type of break did the cylinders have? Not every testing laboratory indicates the type of cylinder break. If the test report does indicate the type of break, as shown in Figure 4-1, something may be learned that could explain the low strength.

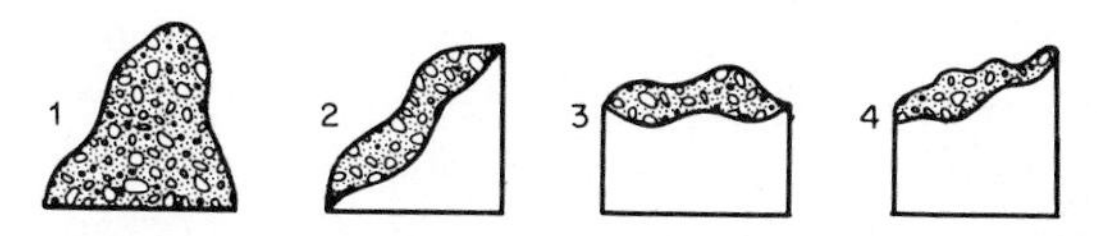

Figure 4-1 Types of cylinder breaks.

In a normal cylinder break under compression, the sides of the sample become almost barrel-shaped the instant before they destruct, leaving an hourglass shape, as shown in Type No. 1 of Figure 4-1. The Type No. 2 break, a shear break, may well indicate unlevel capping. Type No. 3 in Figure 4-1 is typical of poor compaction, usually caused by the failure of one layer of the sample to bond to the previous layer with the tamping rod. Type No. 4 could well be a combination of No. 2 and No. 3.

In addition to the factors that may or may not have appeared on a test report, there are many other considerations. The competence of the individual who molded the cylinders may be questioned. If previous reports on the same job have been acceptable, one may assume that the fault of the low test result lies elsewhere. There is always the possibility, of course, that a different individual instead of the one who cast the previous samples had cast the cylinders on the day in question. Even if the cylinders were molded correctly, it is possible that after being made they were placed in a spot different from the usual one and that some unusual conditions in that area, such as vibration, may have harmed the samples.

Further, it should be determined whether the slump was carefully taken or only eyeballed (most people are poor judges of guessing slump accurately) and from what part of the load the slump and cylinders were taken. Before a sample is taken, it should be made certain that the concrete batch was uniform in its discharge. Moreover, the sample should never be taken from the first or last two yards of the load.

It is necessary to perform both the slump and the cylinder tests from the same sample. This is best done by obtaining the sample in a premoistened wheelbarrow and wheeling it away from the immediate vicinity of the concrete truck, whose ground vibration may affect the slump. Running back and forth with a shovel full of concrete, taken from the truck discharge chute to perform the tests, is poor practice. Keep in mind it is necessary to perform all sampling within 15 minutes.

THE READY-MIX TRUCK DRIVER

Certainly it is impractical for the concrete supplier to be present on every job every day when tests are conducted. But the supplier does have a company representative present during sampling—the ready-mix truck driver. Drivers of concrete delivery trucks are a valuable source of helpful information if trained in what to look for in sampling procedures. A short instruction course for these drivers can pay big dividends to the supplier. Have the drivers note any improper practices and report this information to the dispatcher or the head office. All pertinent data reported should be written and filed under the customer's name for possible future reference.

Some protection for newly made concrete cylinders should be provided in the field, either by the testing laboratory or by the contractor. An insulated storage box, large enough to store a 3-day supply of freshly made cylinders, is generally adequate. The cylinders can be placed there when they are made, or after they have set initially. The molds should have caps to prevent evaporation of water from the concrete. If there are no caps, wet burlap can be placed over the cylinder molds. In cold weather one or two 100-watt bulbs in the storage box are usually sufficient to protect the cylinders from freezing.

The proper procedure for taking a slump test should be demonstrated and carefully explained in the instruction course. The actual slump test should be conducted as quickly as possible to avoid slump loss. Some admixtures, a rich mix, or high-temperature concrete lose slump rapidly due to moisture evaporation.

CONDUCTING THE SLUMP TEST

The clean slump cone should be placed on a level, sturdy platform, and with the test sample within easy reach, the test should be conducted as follows.

Standing firmly on the foot brackets at the bottom of the cone, add concrete to approximately one-third of the volume of the cone and rod the material 25 times with a bullet-nosed steel rod of ⅝-in. diameter (Figure 4-2).

Fill the cone two-thirds full and rod 25 times. The rod should penetrate into the first layer but not go entirely through it (Figure 4-3).

Figure 4-2 **Figure 4-3**

Fill the cone to overflowing and rod 25 times, again penetrating the layer below but not going through it (Figure 4-4).

Using the steel rod, saw off the excess concrete remaining on the top of the cone and carefully clean away the overflow material at the base of the cone mold (Figure 4-5).

Figure 4-4

Figure 4-5

With hands on the upper handles, press down firmly and remove the feet. Raise the cone in a steady upward lift, carefully, but not too slowly. The concrete within the mold will slump (Figure 4-6).

Invert the cone (narrow end down), and place it beside, but not touching, the sample. Place the tamping rod across the top to extend over

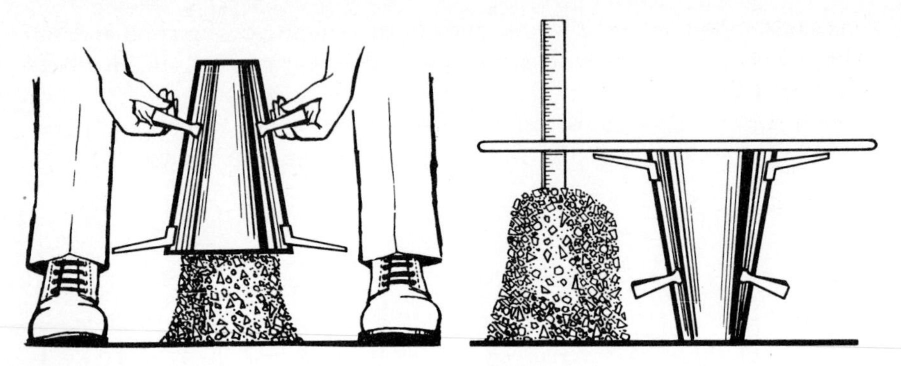

Figure 4-6 **Figure 4-7**

the concrete. The distance from the average top of the sample to the bottom of the rod, measured in ¼ in., is the slump of the concrete (Figure 4-7).

CASTING THE CONCRETE CYLINDER

As soon as the slump test is completed, the concrete from the same wheelbarrow sample, not including the slump-test concrete, should be used for casting the cylinders. Choose a level spot and place the cylinder molds on a wooden platform on this level area. It is then usually necessary to remix the sample concrete, by shovel, as some bleeding and minor segregation will probably have occurred.

The cylinders are then cast in three level layers of concrete, each rodded 25 times as in the slump test, and the top of the molds leveled with the tamping rod in a sawing motion. The more level the concrete is at the top of the cylinder, the easier it will be to cap the cylinder for the compression test in the laboratory. The final act in the cylinder casting procedure is to cover the molded samples to protect them from possible evaporation. Plastic caps are often provided with the cylinder mold for this purpose. If these are not available, cover the samples with polyethylene or wet burlap.

Driver instruction on just these two most frequently conducted field-test procedures offers possible clues to low strength reports.

LOW STRENGTH RESPONSIBILITY

As previously stated, when low-strength concrete is reported, the finger of suspicion usually points at the concrete supplier. The fault may well have been due to concrete as delivered; but what about the testing laboratory chosen to police the project? In some cases the architect or engineer may select the testing laboratory. In most cases the project specifications require the general contractor to retain a testing facility. In such cases the contractor will often choose the laboratory on the basis of price, which is not always in the best interest of all concerned.

TESTING AND INSPECTION

Testing and inspection are good insurance. On an important project they are a must. Since competent inspection protects all concerned, it seems logical to select a reputable laboratory company and not be concerned about the few extra dollars spent. It is beneficial to have the structural designer, rather than the contractor, choose the testing agent. A job inspector from a laboratory retained by the contractor may be inhibited from being as strict as he might normally be.

When a testing laboratory is chosen, it would be wise for all concerned to visit the laboratory and check the facilities. Has the testing machine been checked for accuracy recently? Is the moist curing room operating with the proper temperature controls? Is it large enough to store sufficient test cylinders at the height of the busy season? Do the cylinders on the stripping table, or nearby, indicate they are to be tested on the date of the visit or on the following day? Just getting the cylinders to the laboratory is only a part of proper procedure. A visit to the laboratory may prevent a lot of headaches later on. If the testing agency is reputable and competent, the owners will be proud of their facilities and only too happy to show you around.

CONCRETE CORES

However, in spite of the best precautions, the cause of a low strength report may not be explainable. When no reason for the low strength is apparent, cores may be demanded. Several research agencies have conducted tests over the years comparing the strength of concrete as delivered by the supplier against the strength of cores taken from a section of the structure where the delivered concrete was placed. In 90 percent of these comparison tests the cores failed to match the strength of the cylinders. Core strengths of 30 percent and lower were not unusual.

Tests have also shown that cores taken from slabs come closer in strength to molded cylinders of the same concrete than do cores taken from columns and walls. This corroborates the theory that slabs are more easily and generally better cured. Of further interest is that concrete cores taken from the lower sections of columns and walls have higher test strengths than cores from upper sections, probably because of the

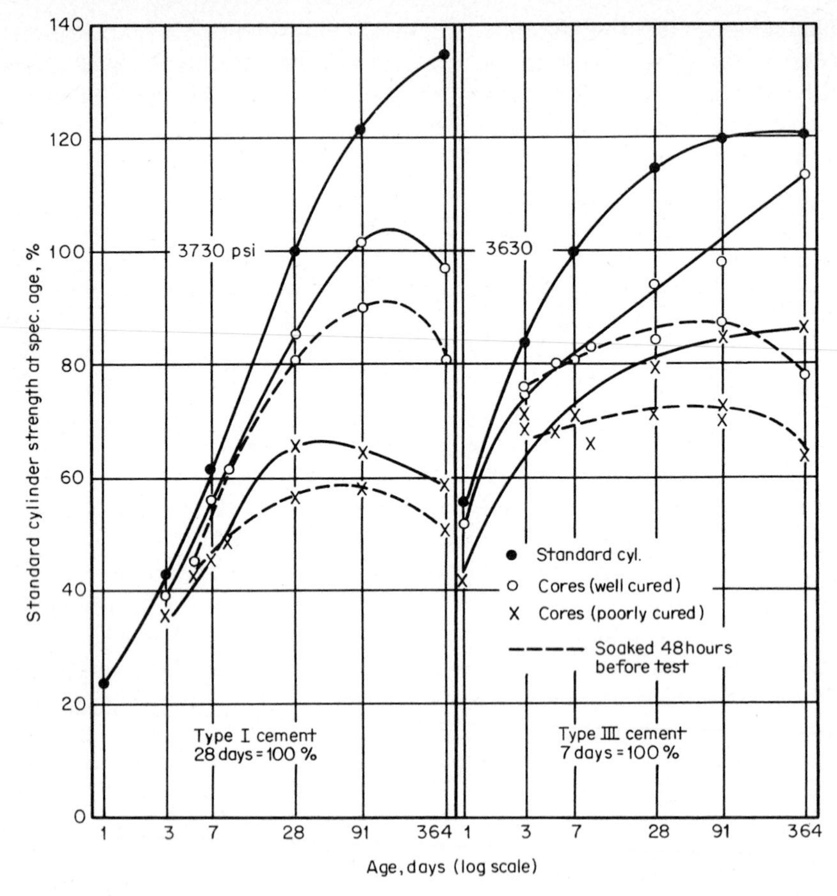

Figure 4-8 Chart showing variations of the strength of standard cylinders versus cores taken from same concrete, including the loss of strength due to lack of sufficient curing. (Source: National Ready Mixed Concrete Association.)

better compaction of lower concrete from the weight of the concrete above.

Lightweight concrete cores compare more favorably with standard cylinder results than do those with other aggregates, probably because of the higher water content within the aggregate acting as a curing agent.

Concrete cylinders tested in comparison test higher when dry than when wet. This same fact is true with cores. Conversely, concrete beams tested for flexural strength test higher when wet than dry. One should never use the Swiss hammer, or other impact test methods, in the field when the concrete is moist or wet.

In light of the above facts, it is suggested that the concrete supplier have an understanding with the general contractor that the supplier is responsible only for the strength of the concrete mix as delivered from the truck at the point of discharge and tested in the standard manner.

TEMPERATURE: EFFECT ON CONCRETE STRENGTH

Even if all testing procedures have been followed correctly, tests may still show low strength results. Again, this may not be the fault of the concrete as delivered, but it could result from weather conditions which existed on the day the samples were taken. It could have been a hot day, and the truck from which the sample was taken could have been delayed in discharging the load. Rising concrete temperatures play havoc with air contents, causing rapid slump loss. Frequent additions of water to maintain slump during the delay would not reflect the true strength of the concrete as originally delivered.

Figure 4-9 is clear evidence of the high water demand necessary to maintain slump as the concrete temperature rises. The slump might have been maintained, but W/C ratio was drastically increased, resulting in lower-than-required strength. Even under normal weather conditions, high-temperature concrete can occur with an overly rich mix (a high cement factor), or can occur because of long hauling distances from the concrete plant to the jobsite.

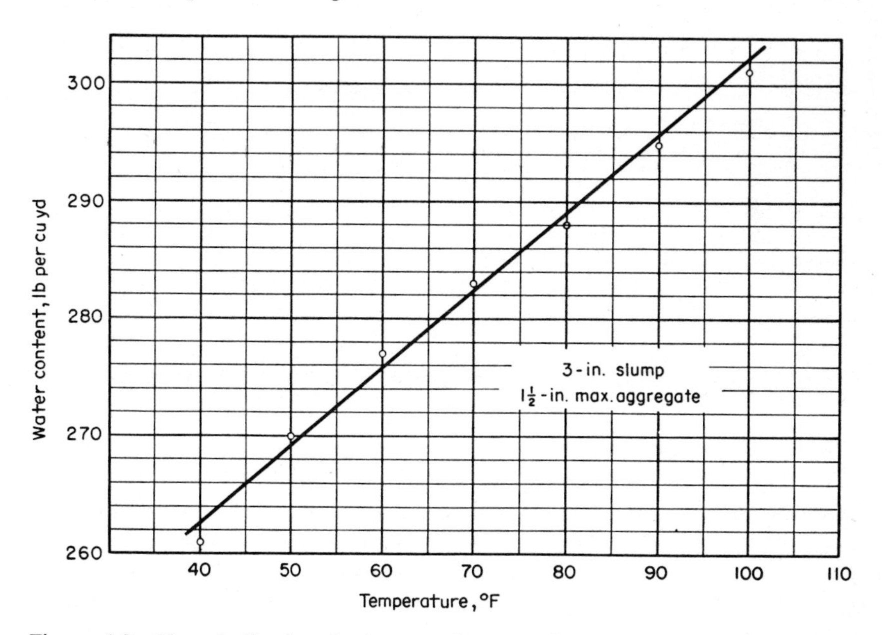

Figure 4-9 Chart indicating the increase in water demand of a concrete mix with an increase in temperature. (Source: National Ready Mixed Concrete Association.)

It can be noted that concrete curbs that appear in good condition at the beginning of the concrete placement frequently show progressive signs of deterioration as the placement proceeds, until, at the end of the delivery, there is complete disintegration. Because the curb forms are narrow, the placement of the concrete is a slow process. The time involved in discharging the load can increase the concrete temperature, reducing the air percentage and requiring frequent additions of water to maintain slump. Strength would be lower than specified, and the lack of proper air content would result in low resistance to wear and no resistance to freezing and thawing. (See Figure 4-10.)

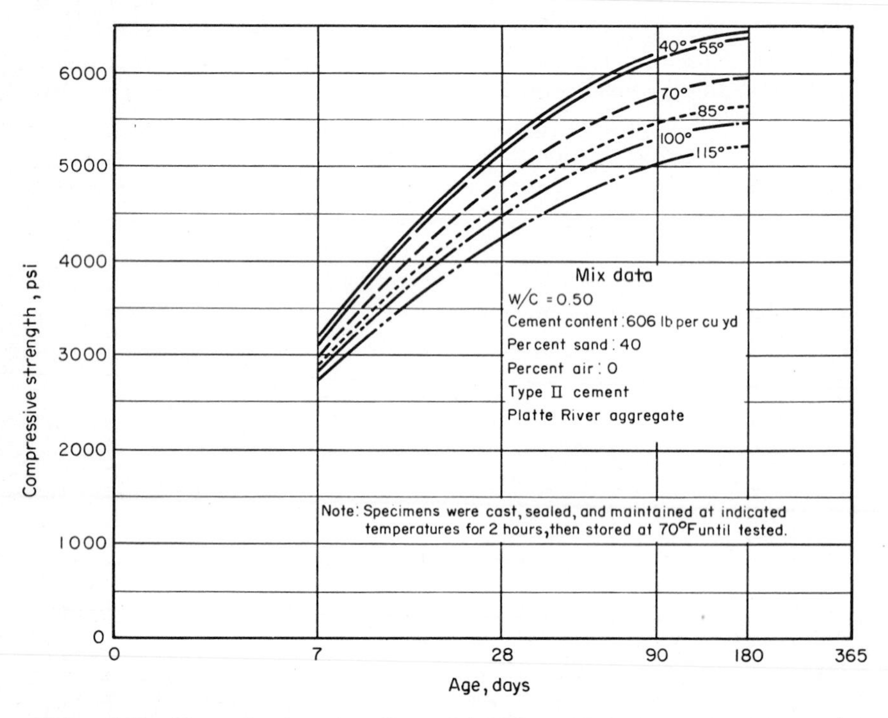

Figure 4-10 Chart showing the effect of initial temperature on the compressive strength of concrete. (Source: National Ready Mixed Concrete Association.)

There is no question that high initial concrete temperatures are detrimental to quality concrete. Although there is little to be done to control weather, the contractor can try to eliminate any unnecessary delays at the construction site. He should not request more concrete delivery trucks than can be handled quickly and efficiently. If a truck has to stand in line with its barrel turning, the concrete temperature will rise and create problems.

Under hot weather conditions, the concrete supplier can spray his coarse aggregate stockpile with a *fine* spray from a garden hose to reduce the aggregate temperature by evaporation. Discharging all mixing water from the holding tank of the concrete truck after each round and refilling with fresh water will help keep the concrete temperature down. Use the water in the tank returned to the plant to clean and cool the mixing barrel. Insulating the water tank and painting the mixer a light color is a *definite plus*. If all these precautions fail to lower the concrete temperature, the addition of shaved ice may be the only solution to reduce the concrete temperature to an acceptable level.

Most *concrete* specifications require that concrete with temperatures above 90°F be rejected. *Cement* temperatures above 180°F should be avoided, but hot cement alone will cause only a minimal increase in concrete temperature. The use of Type I or Type II cement has little effect on concrete temperature rise. However, there are many concrete producers who prefer the use of Type I cement in the cooler weather and Type II in the warmer months. Their reasoning for the change is that Type II cement, being a modified retarder, allows more finishing time and reduces the possibility of a quick set.

THE CONCRETE SUPPLIER

So far in this chapter the various factors that can contribute to concrete strength loss have been discussed on the assumption that the concrete supplier may not have been at fault. If the construction project has been underway for some time with satisfactory strength reports, this assumption is reasonably logical. There is, however, the possibility that the concrete, as delivered, created a problem.

Even though the concrete plant may be equipped with a digital recorder, showing the date, truck number, point of delivery, and scale weight, proving all the ingredients of the batch were weighed properly, the supplier is not entirely above suspicion. If a different type of aggregate had been inadvertently used, or if the gradation of the aggregate(s) had changed drastically, the effect on compressive strength would be detrimental. Further, if the aggregate contained organic impurities, because of inadequate washing, lower strengths could result. In such cases it would be hoped that the testing laboratory might examine the failing cylinders for a possible explanation of the low test(s).

When a test cylinder reflects much lower strength than normal, most test labs will examine the failing cylinder for a possible clue to its poor performance. It is good practice and can often detect the cause of the failure.

Where the plant is not equipped with an automatic recording device, there is always the possibility that the weigh master erred in batching.

The weigh master is human and can make mistakes, but generally this is the exception to the rule and should be an isolated case (if it happens at all). However, it can and does happen, particularly when the regular weigh master is sick or on vacation, and the replacement is the batcher the day the problem occurs.

It is a good practice, and one usually followed, to have the weigh scales checked for accuracy every 90 days. However, should they go awry in the interim, there would be a noticeable difference in the appearance of the concrete, and the need for corrective action would be obvious.

There is also the possibility that the concrete delivery truck had a breakdown on the way to the job. A long delay could require frequent additions of water, thereby increasing W/C ratio. Unfortunately, it is difficult, and sometimes impossible, to fully explain the cause of an occasional low-strength cylinder or group of cylinders. There are so many possible causes to be considered 28 days after the delivery date that there is often no single plausible explanation that can be isolated. The problem of low-strength concrete is often not created by one single item but by several minor infractions of good concrete practices.

UNUSUAL CIRCUMSTANCES CREATING LOW STRENGTH

Many stories have been told by people in the industry of unusual circumstances creating low-strength cylinders from acceptable-quality concrete. One case in particular had an interesting twist. The project was an important one, and from the very first pour the tested cylinders fell far below required strengths. The supplier had an excellent reputation and was completely frustrated by the poor reports. His materials were checked and his scales were tested, and even when he increased the cement factor of his own volition, the cylinders failed. The testing laboratory was well known for its high-quality personnel. Cylinders were carefully molded and placed well off the jobsite, carefully protected, and carefully carried from the site to the laboratory; the lab procedures followed. After many frustrating delays for everyone concerned, it was discovered that the shed where the cylinders were being job-cured was above a subway train route, subjecting the cylinders to almost constant vibration.

Two sets of cylinders were sampled from the same batch to verify the cause of the low-strength concrete. One set was placed in the shed, the other on another part of the jobsite. The shed samples failed as usual, while the other cylinders broke well in excess of specifications.

5/FRESH CONCRETE

PLASTIC QUALITIES

Fresh concrete is the plastic concrete as discharged from the mixer or as discharged from the ready-mix delivery truck. Fresh concrete should be easy to mix, transport, place, and finish, and should remain free from segregation. Several conditions are required, in combination, to produce these quality features.

1. Enough cement and water paste to fill the aggregate voids and hold the aggregates in suspension
2. Good gradation of the fine and coarse aggregate
3. The proper percentage of the fine to total aggregate
4. The proper shape and surface character of the aggregate particles

WORKABILITY

When all the above features are in proportion, the plastic concrete will have good workability. Workability should not be confused with consistency; the term *consistency* refers to the wetness or slump of the mix. Excellent workability, or flow, can be better attained with a lower slump in well-designed concrete containing well-graded aggregates than it can at higher slumps in concrete having poorer-quality materials.

Unfortunately, the term *workability* is difficult to explain. With enough experience, one can rely on visual inspection alone to define the concept and determine the workability of a batch. The concrete should be homogeneous at the medium slump range, be easy to trowel, and not have water gain (bleed) rise quickly to the surface. Even where good-quality aggregates are used, with a sufficient water and cement paste, the concrete will lose workability when the S/A ratio is out of proportion.

The average concrete producer delivers his product to many different jobsites in the course of a day. Many customers are do-it-yourselfers, or amateur masons undertaking a great variety of types of work. Customer acceptability, by such a diversity of users, is of prime importance in the design of the delivered concrete. The producer, as a result, must design his concrete to meet strength requirements, to be both durable and workable, and if possible, not to segregate under high-slump conditions.

All too frequently, the nonprofessional mason will demand "more sand" when the mix, as originally designed, has good workability characteristics. With reluctance, the supplier, fearing the loss of the customer on future work, may adjust his aggregate proportions to comply.

On the smaller jobs, quality control is usually nonexistent, with excessive slumps frequent and handling and finishing accomplished haphazardly. Along with these practices the concrete supplier has to contend with variations in temperature; with wind, rain, heat, and cold; with slow pours; and frequently with insufficient time to mix the load properly.

On large projects the concrete mix is generally designed by a testing laboratory to conform to specifications. Slumps are controlled, and the placing and finishing of the concrete is inspected. As a result, problems are at a minimum. Further, the masons employed on these larger projects are professionals whose expertise contributes largely to the success of the pour.

BLEEDING

The consistency or slump of the concrete is as important as the workability. The specific gravity of the aggregates is 2 to 3 times heavier than water. In high-slump concrete these aggregate particles, no longer in suspension, settle in the mass, displacing the water and channeling to the surface to create bleeding. How long this bleed water remains on the surface depends largely on weather conditions. On damp, cloudy, windless days, it can remain for hours.

The longer the water remains on the surface, diluting the paste, the less durable the wearing surface of that concrete becomes.

What the maximum slump is before segregation and excessive bleeding takes place depends largely on the design of the mix and the quality of the materials used. Generally it can be assumed that any concrete with a slump of 6 in. or more, is excessive. Most concrete mixes can be handled easily at a 4-in. slump; a 5-in. slump is sufficient for good placing and finishing qualities.

The surface problem of excessive bleeding invariably causes the concrete below the surface to suffer. The water, channeling its way to the surface, weakens the cement paste on the bottom of the coarse aggregate and on reinforcing bars, reducing bond. Further, the channeling brings air-entrained bubbles to the surface, where they then dissipate into the air, and do little good for the concrete. Finally, these channels offer perfect tunnels for the entry of moisture at later dates to cause deterioration within the concrete.

Some bleeding of concrete placed in normal slump ranges is to be expected. After concrete has been placed in a form and leveled off, and before final set has taken place, it is often observed that the top surface has settled below its original level.

SETTING TIME (VIBRATION LIMIT)

Fresh concrete gradually stiffens until it has reached its vibration limit: it is then considered, for all intents and purposes, to have set. The time it takes to reach the set condition will vary with the cement factor of the mix, the temperature of the concrete as placed, the type and brand of cement, and weather conditions. The temperature of the fresh concrete is strongly affected by temperatures below 50°F, which cause a delayed set.

When the weather starts getting cold, hot water is usually added to the mix to keep concrete temperatures well above the 50°F mark. When ambient temperatures are below freezing, or nearly so, the aggregates are heated along with the water. However, concrete temperatures above 70°F can quickly drop below 50°F when the concrete is placed on a cold or frozen subgrade.

The use of space heaters or other heating methods during and after the placing of concrete on a frosted subgrade will create internal stresses as the surface sets before the concrete on grade. This improper set can cause *curling,* i.e., a raising of the concrete around the perimeter.

VOLUME CHANGE (SHRINKAGE)

Concrete placed under normal conditions will undergo some volume change. Concrete shrinks as it dries largely because of a reduction of water from evaporation and from the utilization of the water to the cement paste. The greater the amount of water in the fresh concrete, the greater the shrinkage. Most concrete slabs will evidence some minor cracking because of this shrinkage. The higher the slump, the greater the probability of increased shrinkage cracks.

HIGH-SLUMP CONCRETE

Naturally, drivers will have varying degrees of competence in judging slump. Some seem to take forever to arrive at the desired slump, while others do so quickly and without effort. There are careless drivers who occasionally mix the concrete with slumps far in excess of a customer's wishes. *Such concrete should be rejected.* All too often, for various reasons, the customer will accept the overly wet concrete delivery. This poor-quality concrete is sure to result in weak, porous concrete that will create problems.

AIR ENTRAINMENT

One of the many advantages of air entrainment in concrete is the ability to reduce segregation and bleeding by means of the billions of minute air bubbles which help hold the aggregates in suspension. In high-slump concrete the air is drastically reduced, creating a segregation and excessive bleeding. These two factors alone reduce strength, resistance to freeze/thaw cycles, and durability.

As previously mentioned, high-slump concrete will result in greater volume change and a greater number of shrinkage cracks. Yet, in spite of these facts, the concrete placed on uninspected jobs is, and probably will continue to be, placed with concrete having higher slumps than necessary.

Although air entrainment is credited with being one of the great advancements in concrete technology, there are some areas in which it is not used. In parts of the South, for example, except on state or federal projects, air is not always a part of daily production. Because many suppliers and contractors feel the main purpose of air is to resist freezing and thawing, they do not request it in those areas where freezing temperatures rarely if ever occur.

The use of air entrainment is a plus in concrete in any climate. One of its features, among others, is its ability to help reduce bleeding and segregation of a concrete mix. True, it can reduce the strength of some concrete mixes, but the overall advantages will outweigh the minor reduction in strength. In concrete slumps above 6 in., the valuable qualities of air are greatly reduced.

As long as the practice of using high-slump concrete continues, the cause of many concrete problems will remain.

6/Handling concrete

SITE EXCAVATION

The first subcontractor on a construction site is the excavator. His job is an important one. Proper footings depend on his ability to remove the soil to an accurate depth and to disturb the subgrade as little as possible. All footing subgrades should be thoroughly compacted whether fill is used or not. Too deep an excavation will demand fill, and no matter how well this is compacted, settling of the planned structure can occur and create unwanted cracks. Too high a grade, or an unlevel one, will necessitate leveling the grade by hand, which is time-consuming and costly.

FOOTING FORMS

Footing forms must be set to proper line and grade, and they must be well staked and braced. The footing concrete should be placed with a reasonable slump, to avoid a top scum—called *laitance*—which can result in a poor bond for the foundation walls to come later. It is recommended that the footing be keyed to reduce water seepage. Where the soil offers poor drainage, or on special projects, a water stop is recommended.

WALL FORMS

Before wall forms are erected, the footing of the surface should be clean and free of debris. The wall forms should be treated with form oil *before* erection so that the form oil will not run down the forms to the footing concrete.

Enough cannot be said regarding the necessity for well constructed wall forms, especially since solid form construction is a detail often overlooked by some contractors. Forms must be snug and tight, firm and well braced, to avoid bulging or shifting. The form does not have to give way completely to cause cracks. Any movement of forms due to timber expansion, or to a loosening of nails or clamps, can create unnecessary problems.

Depending on the height of the wall, the concrete may be discharged directly from the concrete truck, be placed by means of buggies or by crane, or be pumped. Regardless of the method used, it is crucial, in order to avoid segregation, that the concrete be placed straight downward and not be allowed to bounce off the wall forms.

Serious segregation can be avoided when using buggies or cranes by depositing the concrete into a *hopper,* or funnel-shaped receptacle, on top of the form. Below, and attached to the hopper, is a series of tapering metal sections known as an "elephant trunk," which can be raised to keep the metal sections above the level of the deposited concrete as it rises in the form.

A concrete pump may be used on high-wall-form construction to avoid concrete segregation. The 4-in. rubber sections at the discharge end of the line can be lowered down into the form, reducing aggregate bounce off the reinforcing steel. Some contractors use windowlike openings located at different levels in the form. When the concrete reaches these levels, the openings are then closed with prefabricated sections. If contractors plan well, there is little delay in placing the concrete and the method does drastically reduce segregation.

It is also necessary on high-wall concrete placement to first introduce several inches of grout on top of the footing at the base of the wall form. The concrete placement should follow promptly, while the grout is still in the plastic state, to eliminate the possibility of honeycombing at the base of the wall.

Do not use high slump concrete when placing wall concrete, as it will usually result in unsightly sand streaking because of the bleed water's rising to the surface between the concrete and the form, thus diluting the surface paste.

Do be certain the footing is clean and free of any debris. *Do* treat the wall forms with good form oil before erecting the wall panels, and let them drain before erection. *Do* use internal vibrators rather than external ones. External vibrators, popular in the manufacture of precast products, will place excessive strain on the wall forms. Compacting the concrete on low walls by rodding with a 2 × 4 may do a fair to reasonably good job but will not work well on higher pours.

VIBRATION

Writing a specification for the type of vibrator to be used in a given situation is challenging, and, for most, extremely difficult. The internal vibrator is essentially a steel tube housing the vibrating element. It is operated either by gasoline engines, by air, or by electric motors. Many experienced contractors contend that the electric motor variety is the best to use if a source of power is convenient. The gasoline type, while acceptable, is a second choice because it requires more servicing and is less convenient. Air motors, while being much lighter in weight than electric motors of the same power output, require an adequate air supply.

It is important to use the internal vibrator properly, in a straight up-and-down motion. Unfortunately, all too frequently it is used almost horizontally to flow the concrete into place. Held too long in any one spot, it will create overvibration and bring an excess of fines to the surface, causing laitance.

Vibrators should be introduced and withdrawn slowly at each location. They should be inserted to the full depth of each new layer of concrete, and slightly into the previous layer, to eliminate a cold joint. The slump of the concrete must be kept in mind when using the vibrator. Low-slump mixes require closer entry of, and slower removal of, the vibrator for best conpaction. With such concrete, the individual layers must necessarily be placed with a thinner layer than is the case for a more plastic mix.

The time limit for removal of forms depends largely on the richness of the mix, the temperature of the concrete, the slump, and the atmospheric conditions. Under ideal conditions 48 hours may be sufficient time; otherwise, it may be advisable to wait 4 days or longer before stripping. Exposing a wall section to the elements before the initial curing period of 3 days can cause undesirable shrinkage cracks.

If rain is predicted when the wall forms are removed, some contractors do not cure the concrete on the assumption that the anticipated rainwater will provide natural curing. Weather predictions are not always accurate, however, and it is best to spray the concrete with a liquid curing agent as soon as the forms have been stripped.

CURING

Curing is one of the most important steps in placing concrete, especially for finished floors, but many people do not fully appreciate the need for good curing practices. The purpose of curing is to prevent the loss of the water present in the concrete mass. Such moisture must be maintained to effect the complete hydration of the cement paste. The greater the amount of water retained within the concrete mass, the higher the ultimate strength of that concrete.

Steps must also be taken to prevent the loss of concrete water to a dry subgrade. Wetting the subgrade thoroughly, but not so much as to create a muddy condition, is a help, but not always sufficient. The use of 4-mil polyethylene gives superior protection from subgrade absorption and also helps in preventing water penetration into the hardened concrete from the subgrade that can occur from hydrostatic pressure.

The ideal way of curing concrete floors is by means of water, either by ponding or by covering the finished concrete with water-saturated burlap, kept moist for at least 4 days. Keep the burlap constantly moistened, particularly when the slab cured under this method is exposed to the elements. Although this method has lost appeal because of the cost and time delay it involves, water-curing will greatly reduce volume change and shrinkage cracks.

No curing method should ever be attempted until the concrete surface has set sufficiently to resist foot traffic. The widely used kraft paper should be placed flush with the concrete surface, be well overlapped, and be held down securely. Polyethylene sheeting resists flush placement against a concrete surface, resulting in a mottled appearance when removed.

When a liquid curing compound is to be applied to a concrete floor, it is important to know what is to be applied to this surface at a later date. Many liquid curing compounds become an integral part of the concrete and prevent adhesion of tile, cementitious materials, or epoxies. Where such applications are to be made later, an inexpensive, short-lived curing compound is a better choice. Further, the choice of the color of the compound should be given consideration. If the concrete will be exposed to the sun the lighter colored compounds should be used to reduce heat absorption.

NONPIGMENTED LIQUID-MEMBRANE COMPOUNDS

Basically there are only two types of nonpigmented liquid-membrane compounds—resin wax and 100 percent resin. When other materials are to be applied to the cured surface, the all-resin type of compound should be used. Normally this type will last for a month or less, and what little residue might remain can be easily removed with a stiff-bristle brush.

The resin-wax type of compound is used primarily on highways and similar surfaces to help resist water and deicing chemical penetration. These liquid sealants, usually epoxy-resin based, can be applied no sooner than a month after the concrete has been placed. They generally require two applications, the first being a light application which permits the solvent to enter the capillary pores of the concrete, and the second being a heavier application which fills the opened pores with resin to reduce liquid penetration of the concrete.

They are applied with rollers or airless sprayers, and they retain their penetration barrier effectively for many years. They have been applied to surfaces of absorbent building materials such as stucco and brick with excellent results. They are fairly expensive but present a great cost saving over other type sealers that may require further applications at more frequent intervals.

COOL WEATHER CONCRETE

It usually takes 3000-lb concrete 28 days to attain that strength under *laboratory* conditions. However, it may not meet that specified strength under *field* conditions. Concrete cylinders tested in the laboratory at 7 days are generally expected to have attained approximately 60 percent of the anticipated 28-day strength result. Again, this is not necessarily the case in the field, especially when the concrete is exposed to cold ambient temperatures.

Most construction specifications do not permit concrete to be placed when the temperature of that concrete is below 50°F. When ambient temperatures are near or below freezing, the placed mass concrete should not be below 60°F, while thin sections should have still higher concrete temperatures.

It is the responsibility of the concrete supplier to produce acceptable concrete temperatures in cold weather. This may require the use of hot water and perhaps the heating of the aggregates. Most producers strive to retain concrete temperatures of about 70°F. On larger pours of several hundred yards, it is difficult to maintain this temperature in the aggregates passing through the holding bins. Fortunately, mass concrete produces its own heat, and lower concrete temperatures have little retarding effect.

The temperature of the delivered concrete is most important, but concrete of 70°F, placed on a cold subgrade, can drop below 50°F in a half hour or less. It is the responsibility of the *contractor* to maintain a warm subgrade temperature *before* the placement of the concrete, as well as to maintain adequate protection of the concrete *after* it has been placed.

It is recommended that in the period from the first frost up until the daily temperature at the jobsite falls below 40°F for more than 1 day, the concrete should be protected for a period of 48 hours. Once the daily temperatures successively fall below 40°F, the fresh concrete should be protected for a longer period.

Concrete generates heat rapidly for the first 3 days after placing and will cure normally if protected from loss of heat. Wall forms protect the concrete to some degree, but additional protection should be provided by canvas drapes, and heating should be provided when bad weather or temperatures are expected to fall below 0°F. The forms should be left in place for as long as possible for the curing and protection of the concrete.

Flat surfaces exposed to freezing temperatures lose heat quickly and should be protected as soon as the concrete can be walked on. Hay, straw, insulating blankets, or other protection must be placed in close contact with the concrete for best results. A polyethylene covering placed over everything provides added protection in case of inclement weather.

If concrete is placed above grade, salamanders or other heating equipment should be placed below the concrete used. It is important to maintain a reasonable temperature in these areas below the concrete being protected. Too much heat can cook the concrete and result in problems almost as bad as if the concrete were attacked by frost. There are thermometers that can record the heat ranges during the heating process. They should be used, and a record of their ranges should be kept as part of the project.

However, when cold weather concrete should be protected and for how long depend upon ambient temperatures, the richness of the mix, wind conditions, and the type of cement used. Earlier strength gains

Figure 6-1 One method of protecting concrete or masonry walls in cold weather. The inside should be heated to a minimum of 60°F. (Source: Portland Cement Association.)

Figure 6-2 Tarp covering for protecting concrete floors and columns in cold weather with proper heating. (Source: Portland Cement Association.)

can be achieved with the use of calcium chloride, or with the addition of one bag of cement if the concrete mix is not overly rich.

HOT WEATHER CONCRETE

An ambient temperature of from 65 to 75°F is ideal for the setting and curing of concrete. When temperatures rise above 85°F, it becomes more important to take steps to prevent evaporation of the mixing water and to reduce quick setting, which would result in finishing difficulties. As previously stated, the temperature of the concrete is important and tends to be higher in hot weather in spite of cooling measures taken by the concrete supplier.

The contractor should plan his pour carefully and have sufficient personnel to handle the concrete promptly, as soon as it arrives. Any delay in discharging the concrete from the truck should be avoided. The friction of the mixing action increases the concrete temperature during prolonged mixing, creating slump loss and a reduction in air content. The air may never be recovered, and water added to maintain slump will produce lower-quality concrete.

Many concrete problems with concrete in the hardened state are created in the placing and handling of the plastic mix. Settlement cracks in walls usually start at the footing as a result of unstable subgrade, poorly compacted concrete, or lack of reinforcement. Failure to install the footing well below the frost line in the area of construction can induce frost heave and result in building cracks.

CONTROL JOINTS

Control joints are placed in walls and floors to prevent random cracking and to predetermine the location of possible cracks if they occur. The control joint is made with various depths, depending largely on the thickness of the concrete, and is made during the finishing operation or cut later with an electric or gasoline-driven diamond saw blade. This saw blade should not be used for at least 12 hours after the concrete has hardened, but should not be used much later than that either.

The wall joint (or notch) in Figure 6-3 is nailed to the form before

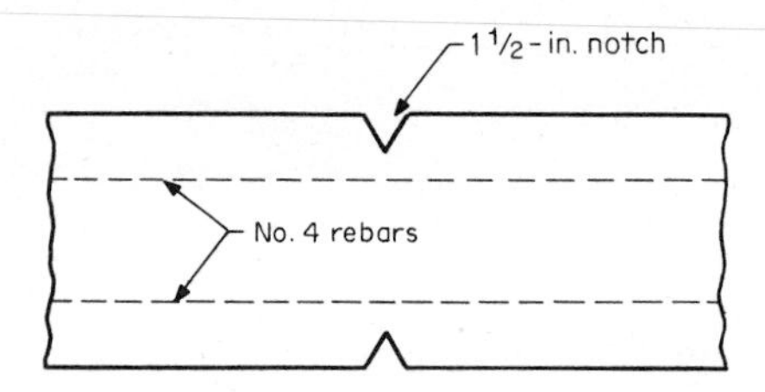

Figure 6-3 A control joint in a foundation wall, spaced every 30 to 40 ft, to prevent random cracking.

the concrete is placed. Should a crack appear at these notched joints, they can be easily repaired with hydraulic cement or similar material. The joints should be installed at a minimum of 30 ft for best results.

Concrete shrinks and expands continually and will produce unsightly random cracking unless control joints, isolation joints, or relief joints are used.

KEY JOINTS

Large floor pours are best left to the experienced mason who has the proper tools and knows how to use them. As shown in Figure 6-4, he will install keyed joints, if necessary, as a form to separate one floor pour from the next pour. He has long aluminum screeds to level the plastic concrete as he proceeds in the placement. He constantly checks the level of the concrete with a transit or laser beam as he proceeds. He further levels the concrete behind the screed with a magnesium bull float to eliminate minor depressions or high points.

The experienced mason knows exactly when the plastic concrete is ready for the power trowel and knows what trowels to use on finishing. He also knows what angle the power blade must have to accomplish the desired result. It is recommended that the larger pours be handled by competent professionals.

Smaller concrete projects may be attempted by any amateur mason,

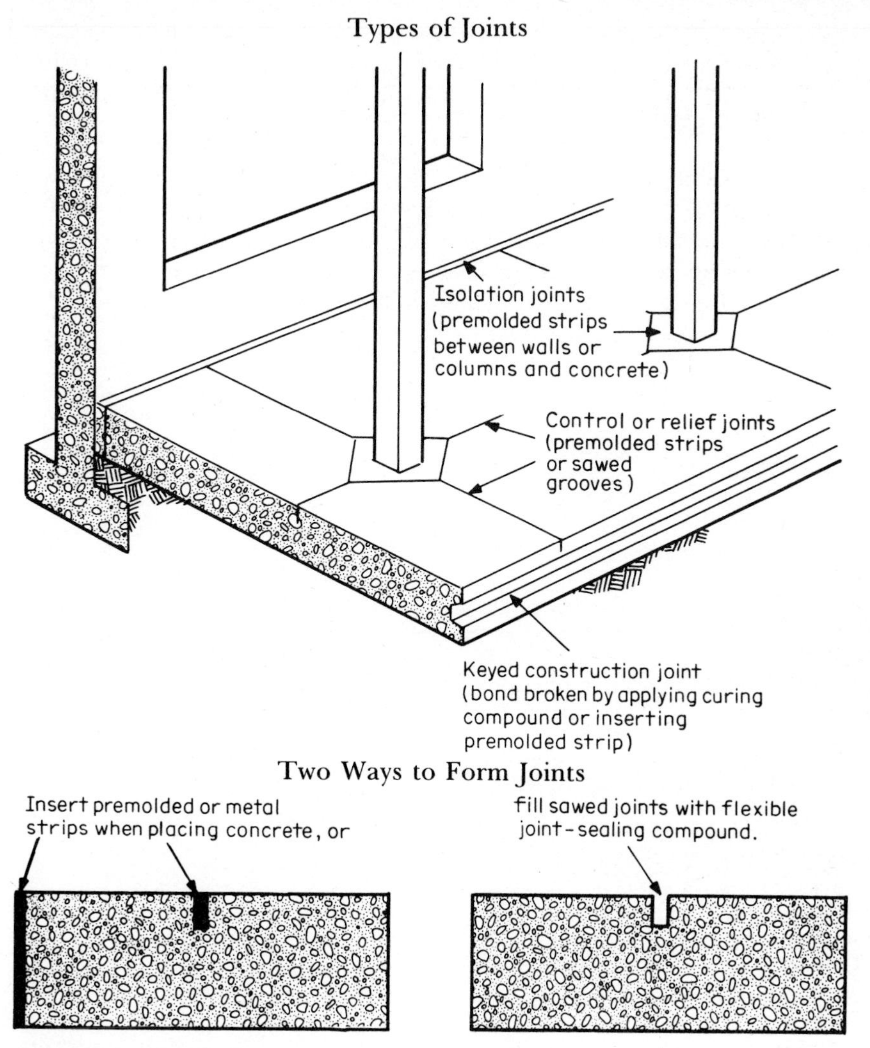

Figure 6-4 Examples of different types of joints and where they are placed for elimination of floor cracks.

and inexpensive booklets giving step-by-step instructions on how to proceed are available in any hardware store or lumberyard. However, even projects such as sidewalks and patios need careful subgrade preparation that takes into account soil and natural drainage.

Unless you are handy with tools and can follow instructions carefully, it is suggested you employ a local concrete mason to do the job. A list of masons can be found in the Yellow Pages of the telephone directory. Obtain prices from two or three masons before making a final choice. Price should not be the sole, or final, criterion in the selection of the contractor.

7 / HARDENED CONCRETE

Many of the problems evident in hardened concrete are created in the forming, placing, and finishing of the concrete in the plastic stage. Some flaws are noticeable during, or shortly after, the initial set; others appear within a few days after placement. Then, there are problems that do not arise until long after the concrete has hardened.

Concrete with low air entrainment, or no air, may not spall on the surface until it has been subjected to freezing and thawing. Overtroweling or overvibration can bring an excess of fines to the concrete surface and cause the concrete to break down later under traffic. Troweling while bleed water is still present on the surface can cause dusting. A great number of cracks in hardened concrete are caused by the shifting of forms from settlement, early load stress, cold joints, lack of early curing, expansion of rust on steel reinforcement, poorly compacted subgrade, alkali-reactive aggregates, and many other factors.

VOLUME CHANGE: SHRINKAGE

In the plastic state the volume of the concrete is reduced in the first hour or two by settlement of the solids. Some water is absorbed by the cement grains, some is absorbed by the subgrade, and some bleed water is lost from the surface by evaporation. All these factors contribute to a reduction in volume. This is the first of many volume changes to occur over the life span of the concrete.

When concrete dries, it shrinks. When it is wetted, it expands. When it freezes, it expands, and when it thaws, it shrinks. Contraction and expansion also occur with changes in temperature. Some volume changes have varying effects. On a free-floating slab a volume change does not create unusual stresses, whereas on concrete restrained by reinforcement, or by perimeter foundation walls, it can create enough stress to cause a great many hardened concrete problems.

Usually little shrinkage will occur if the concrete has been water-cured, unless the plastic mix was placed with excessive slump. When the concrete has been liquid-membrane-cured, some shrinkage is noticeable. But where the concrete has received no surface protection, the shrinkage is quite apparent.

The greater the shrinkage, the greater the possibility of shrinkage cracks. Because cracks spoil the appearance of the concrete surface, a good contractor takes pains to avoid them as much as possible.

Well-graded, sound aggregates will produce less expansion and contraction of concrete than those of low quality. Very rich mixes, with the same aggregates and consistency, show greater volume change than mixes of leaner concretes. Similar tests indicated that the amount of shrinkage also varied widely among different cement brands and types of cement. We must realize that not all cement brands are the same. The characteristics of the raw materials from which cement is made vary in different parts of the country. In general, the finer the cement, the greater the expansion under *moist* conditions. Under *dry* conditions these same cements are not appreciably affected.

Water-reducing admixtures either help reduce shrinkage or have little effect on that property. Calcium chloride used in a 2% solution may increase shrinkage considerably. Entrained air in normal ranges does not affect drying shrinkage to any great extent.

Although the results of tests under controlled conditions are significant, test samples cannot be exposed to the severe conditions of jobsite concrete, where changing temperatures, sun, wind, variations in humidity, artificial heat in buildings, or sudden rainstorms are likely. In spite of all precautions made by the contractor in placing the concrete, the concrete may show faults in the hardened state.

LOAD STRESSES

Concrete can withstand compressive stresses far better than it can withstand tensile stresses. Exposing concrete to load stresses before it has gained sufficient strength will create cracks that reduce the ability of a structure to carry its designed load. The durability of the concrete is further lessened by the entry of moisture through these cracks, corroding the steel, leaching out soluble components, and causing further deterioration of the concrete.

CRACKS: TYPES

A well-known concrete producer with many plants, a private testing laboratory, and a knowledgeable technical staff was asked by a contractor if the producer could guarantee his concrete would not crack. He answered that he could not, but could guarantee that some cracking was very probable. Various types of cracks may occur.

Drying Shrinkage Cracks

Drying shrinkage cracks develop about the time the water sheen disappears from the surface of the concrete. They are usually random, straight, hairline cracks that extend to the perimeter of the slab. Figure 7-1 shows

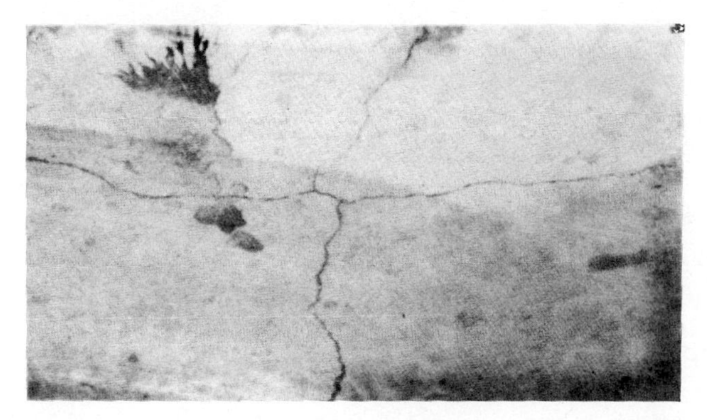

Figure 7-1 Typical shrinkage crack. (Source: *Concrete Construction.*)

a crowfoot shrinkage crack, which generally occurs in the interior surface of a slab and rarely extends to the edge. These cracks are shallow and offer no serious problem beyond marring the appearance of the concrete.

Plastic Shrinkage Cracks

Figure 7-2 shows typical plastic shrinkage cracks. They are wider than drying shrinkage cracks and often extend through the depth of the concrete. They occur more frequently on dry, windy days, emerging parallel to one another and perpendicular to the wind. They are caused by the evaporation of water from the surface and must be repaired as quickly as surface traffic permits. A fine spray from a garden hose, if applied to such cracks as they appear on the surface, will either close them or at least reduce their penetration.

Figure 7-2 Plastic shrinkage crack. (Source: *Concrete Construction.*)

Shifting Form Cracks

Solid form construction, a detail often overlooked by contractors, is necessary for the prevention or minimization of cracks. The form does not necessarily have to give way completely to cause a cracking problem. Any shifting of the form due to timber expansion, or to a loosening of nails or clamps, can induce cracks. Cracks have no particular pattern and can appear during the curing stage or later.

A shifting form crack (Figure 7-3) is another port of entry for moisture to enter concrete and hasten deterioration. Sometimes this crack appears near the form, and at other times further away.

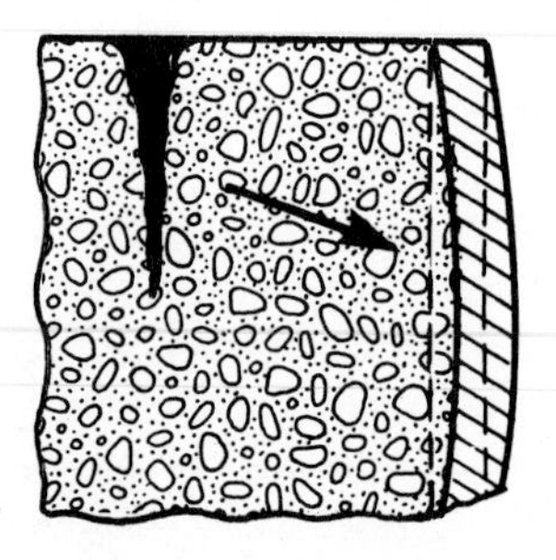

Figure 7-3 Shifting form crack.

Reinforcing Steel Cracks

When concrete settles over reinforcing steel, cracks can appear on the concrete surface above the bar (Figure 7-4). These can be eliminated by placing the bar at a minimum of 1½ in. below the surface and by maintaining a reasonably low slump.

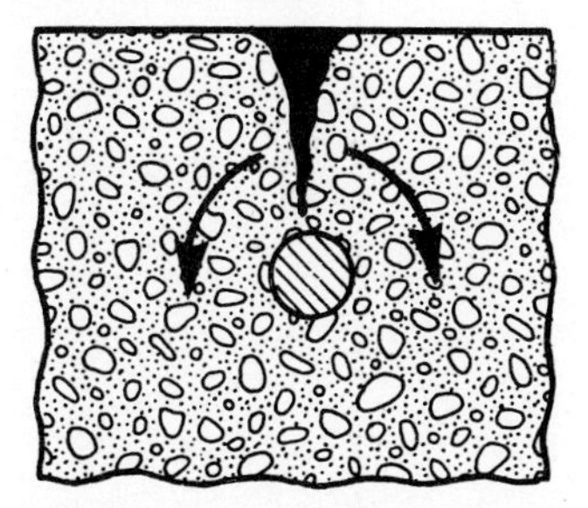

Figure 7-4 Concrete settlement over steel.

Rusted Steel Reinforcement Cracks

Rusted steel reinforcement will expand and create surface cracks. These cracks (Figure 7-5), unless carefully repaired, will permit entry of moisture, creating more rust, and will promote inner-concrete deterioration.

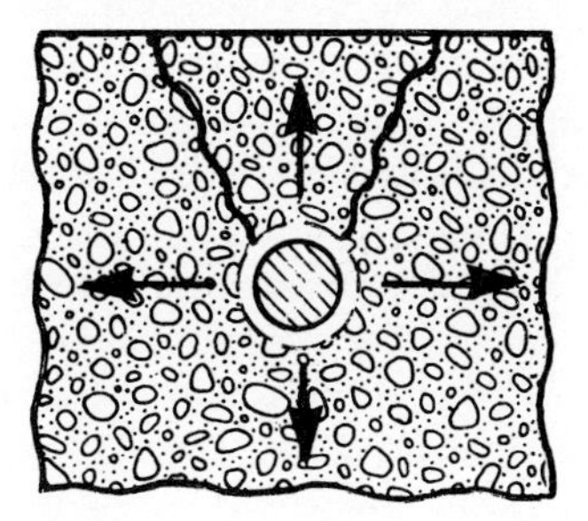

Figure 7-5 Rusted steel cracks.

Subgrade Paper Rupture Cracks

Subgrade paper rupture, or poorly compacted subgrade, allows concrete to shift while it is setting, thus creating cracks (Figure 7-6). There is no particular pattern to the cracks.

When cracks appear in concrete weeks or even months after the concrete has been placed, it is difficult and sometimes impossible to determine the probable cause. Was the subgrade thoroughly compacted? Were the forms well constructed? Was the concrete placed and vibrated in accordance with good construction practices? What was the rate of placement? What were the weather conditions on the day of the concrete placement, and shortly after? What was the slump, the air content, and the concrete temperature? Some of these facts which may be helpful in determining the cause of a problem are not always available at a later date.

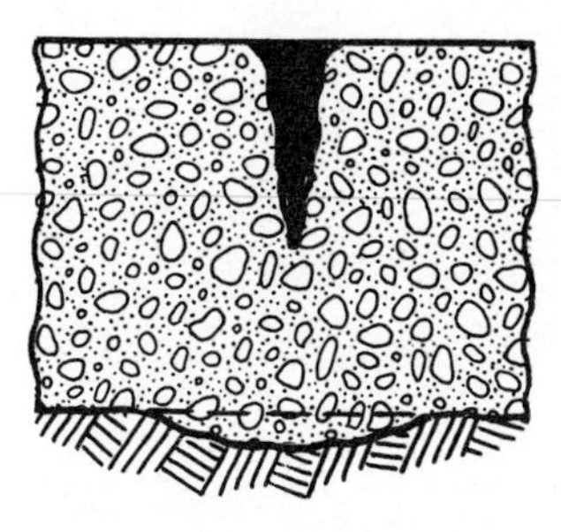

Figure 7-6 Subgrade paper rupture crack.

STEEL REINFORCEMENT

The reinforcement, or lack of it, is critical, particularly in flat work. Alone, concrete may have good compressive strength but poor flexural strength. Concrete must have steel reinforcement if it is to sustain bending loads. It is the steel bars that carry the load in tension, the concrete being merely a filler.

Structural engineers usually design the reinforced concrete to withstand loads in excess of the anticipated stress. However, they are not infallible. The owner of the building to be erected may tell the design engineer the weight of machinery or other material to be stored within the building, but how those loads are to be moved into the building may not be explained. If heavy supplies or equipment are carried across the concrete floor on dollies, subjecting the concrete to loads at the point of the dolly wheels far in excess of design, serious cracking is bound to happen.

When loads are imposed on a concrete floor before it has reached its designed strength, the probability of cracking is heightened. Normally, concrete should not be exposed to load stress for at least a month after placement, and a longer period of time may be necessary if construction was under way during cold weather.

Many concrete projects show little or no serious imperfections for years after completion. Then overnight, it often seems, cracks develop in a structure previously free of such problems. Since concrete is known to grow stronger with age, one can question why a sudden failure should occur so much later. (See "Producer's Responsibility," page 162.)

CONCRETE WEAR

Cracking is only one problem that may show up in older concrete. Concrete floors that have stood up well for a time may begin to show sudden signs of wear. Here again the problem could well have been created during the placing and finishing of the plastic concrete. Frequently, the whole traffic floor area will show signs of wear at once. On other floors the scaled areas may show up in spots.

Troweling too late, after the concrete has set, could be a cause of surface wear. Dusting dry cement on a wet concrete surface to reduce surface bleeding is a possible cause. Frequently, the practice of casting water on a concrete surface, usually with a calcimine brush, may be noted during the finishing process. This procedure will weaken the surface and eventually show up in poor wearability.

A *jitterbug,* a metal grill-type hand tamper, should be used on very low slump concrete to aid in embedding the coarse aggregate and to bring mortar to the surface. If used in concrete without a very low slump, it will bring an excess of fines to the surface, lowering the durability of that concrete. Such a practice, as with overtroweling, will cause surface scaling at a later age.

CONCRETE BUCKLING

Everyone who drives a car is familiar with concrete pavement buckling. In extremely hot weather the concrete, on both sides of an expansion joint, will rise when the amount of thermal expansion of the concrete exceeds the capacity of the joint to control the movement. It can happen to a lesser degree on sidewalks and other exposed areas. This problem is sometimes noticeable on boiler-room floors and in other high-heat enclosures. In either case the concrete on one or both sides of the buckling will result in a trouble spot and eventual cracking.

EXPANSION JOINTS

It is the nature of concrete to crack, and without relief or expansion joints to control them, cracks will appear in unwanted places. High-slump concrete will exhibit more shrinkage cracks than lower-slump concrete. High-slump concrete is porous and weak and unable to withstand the stresses of ordinary volume changes. A well-designed concrete mix with the proper cement factor, and sound well-graded aggregates, will still crack under poor workmanship.

Some contractors are not masons and lack a thorough knowledge of the product they handle. Wire mesh, for example, is used to restrain concrete cracks when or if they occur. For best results this mesh should be placed about 1½ in. below the surface of the concrete. All too often this mesh is laid on the subgrade and the concrete placed above it, where it does little good.

DRAINAGE

Concrete placed on grade requires good drainage to prevent hydrostatic pressure of subgrade water that can produce cracks. To relieve hydrostatic pressure cracks, 4 in. or more of coarse, well-tamped aggregate is necessary above the subgrade, where drainage is poor.

Still, even with good concrete, with good form and subgrade preparation, and even with the proper placing and finishing procedures, cracks can nevertheless appear years later. Figure 7-7 shows a driveway corner

Figure 7-7 Settlement crack. (Source: *Concrete Construction.*)

crack resulting from loss of subgrade. Many driveways and walks are edged to keep grass from spreading over the concrete. This edging, if deeper and wider than necessary, will form a sluice for rainwater to run through, and this in turn will cause the subgrade material to wash out. Frequently the weight of an oil delivery truck or a moving van exerts loads in excess of design and creates cracks. Prompt attention to repair of these cracks is necessary to prevent the cracked section from sinking and breaking free.

Sometimes cracks can be lived with, depending on their nature and on traffic conditions. Complete removal and replacement of the concrete is costly, and often the time delay for such an operation renders it out of the question. Usually a floor topping is considered as a possible solution. The thickness of the topping will depend largely on the load the floor is usually subjected to, as well as on other considerations.

FLOOR TOPPINGS

When floor toppings are to be applied, it is best to select an experienced, reputable contractor. A concrete overlay, when not properly placed, can result in more of a problem than the original floor condition presented.

The old concrete must be water-saturated before the new floor topping is layed. Otherwise the highly absorbent old concrete will steal water from the topping mix, causing lack of bond between the two layers, and resulting in severe cracking and an unsound surface. After the old concrete is well-saturated, but puddle-free, a bonding agent should be applied to assure good bond of both layers. For best results the topping should be applied when the bonding agent is still tacky.

If a concrete topping is being considered for a large area where active cracks and unlevel concrete at expansion or control joints are present, one might consider a different approach to cure the problem.

After the old surface is saturated, dry sand should be spread and leveled over the entire area. Cover the sand with 6-mil polyethylene and top with at least 2 in. of concrete over the sheeting. Control joints should be saw cut at proper intervals the following day, after the concrete has set. Cure with wet burlap and keep moist for at least 4 days.

Several of the more common concrete problems with hardened concrete are described in the following pages.

DISCOLORATION

Discoloration (Figure 7-8) of hardened concrete can be caused by many factors.

1. Dusting dry cement on a wet concrete surface to hasten the finishing operation, particularly if the dry cement is not the same brand and type as that used in the concrete mix.

2. Overtroweling, which causes burned areas.

3. Using different slumps in the same area if there is a wide variance. When colored concrete is used, it is most important to keep the slump consistent.

4. Changing the cement brand in concrete deliveries.
5. Adding calcium chloride in concrete, affecting color by delayed hydration of the ferrites in the cement.
6. Placing concrete on a wet subgrade having puddles of water. This water can leach out through the concrete to the surface, leaving that area much lighter in color than the surrounding concrete.

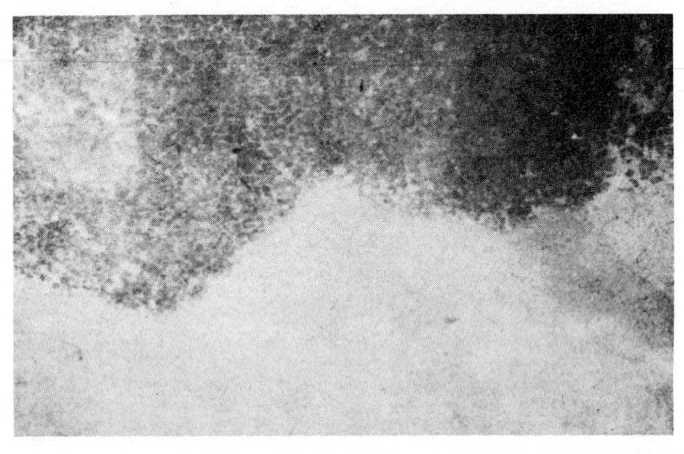

Figure 7-8 Discolored concrete. (Source: Portland Cement Association.)

The use of phosphoric acid, obtained at most chemical supply houses, will generally alleviate most of the causes of discoloration; 1 pt of this acid to 2 gal of water is recommended. Scrub the entire area with this solution using a stiff-bristle brush. Initially, upon treatment, the concrete surface will turn whitish, but this bleaching will gradually disappear.

CRAZING

Concrete crazing on hardened concrete, as shown in Figure 7-9, usually appears shortly after the finishing operation has been completed. Surface shrinkage cracks, usually appearing on low-humidity days when a wind or breeze is blowing across the surface, are formed. They are not a serious problem and occur because the surface dried too rapidly. Some people like the marbleized effect. When it is found to be objectionable, a light grinding will remove the blemish.

SCALING

Scaling (Figure 7-10) is a breakdown of a concrete surface. This surface erosion begins as a flaking off, in several areas, of the top $\frac{1}{16}$ or $\frac{1}{8}$ in.

of the concrete surface. If left alone, this erosion may eventually spread from these areas to an entire section. Listed below are several possible causes of scaling.

1. Overtroweling, which brings an excess of fines to the surface, weakening the bond of the surface to the concrete below

Figure 7-9 Crazing. (Source: Portland Cement Association.)

2. Placing concrete in the fall of the year, when cool temperatures do not allow the concrete to cure properly and gain sufficient strength to withstand subsequent freeze/thaw cycles and the applications of deicing chemicals

Figure 7-10 Scaling. (Source: Portland Cement Association.)

3. Using concrete with low air contents in areas exposed to severe winter weather

4. Using heavy applications of salts and deicing chemicals

5. Using high-slump concrete
6. Using improper curing methods or lack of curing
7. Using insufficient protection of fresh concrete during cold weather temperatures

Strangely, the cement factor appeared to have little effect on the ability of concrete exposed to severe cold temperatures to prevent scaling. When properly placed, finished, and cured, the leaner mixes stood up as well as the richer mixes.

A list of preventive measures to avoid scaling is presented below.

1. Do not use slumps in excess of 4 in.
2. Use air-entrained concrete of 4 to 6 percent.
3. Avoid troweling the surface while bleed water is still present on the concrete surface.
4. Avoid overtroweling.
5. Slope concrete for good drainage.
6. Provide adequate curing.

CURING

Adequate curing refers not only to curing the concrete after it is finished, but additional further protective curing a month after the concrete has been placed. This second treatment is just as important as the first. An application of a 50 percent boiled linseed oil and 50 percent mineral spirit solution, or any of the penetrating sealers, is important in protecting exposed concrete from freezing temperatures.

Fairly recently, epoxy-type resins, referred to as *penetrating sealers,* became available. The product composition includes a solvent to open the surface capillary pores, which are then filled with the synthetic resin. As with the linseed oil application, the concrete surface is sealed, which prevents entry of destructive liquids. The epoxy penetrating sealers are more expensive than the linseed oil mixtures, but they generally pay for their use in providing longer-lasting protection.

8 / Pumping concrete

BENEFITS OF PUMPING

Contractors who have kept accurate cost records can attest to the substantial savings of pumping concrete over the traditional methods previously employed. In the past 10 years the percentage of contractors using concrete pumps has risen from under 5 percent to an estimated 50 percent.

The mobility of the pump machine and the continuous flow of concrete through its line to the site of placement are the principal factors in its widening acceptance. However, it was not until the early pump was improved and more was learned about the modifications of the mix design, necessary when using a pump, that it gained its present popularity.

HISTORY OF PUMPING IN THE UNITED STATES

Figure 8-1 shows a typical concrete pump used in the early nineteen-fifties. After the concrete was delivered into this inverted cone-shaped hopper, it was then drawn into the piston chamber at the base by gravity and by the suction of the piston stroke. The necessity of a plastic mix with such a pump is obvious. The aggregate particle shape, gradation, S/A ratio, slump, and cement content had to be nearly perfect for good pumpability. Blockages were frequent, and harsh mixes caused excessive wear on the liners and on the adjustable valves.

Over the next few years improvements were made in the design of concrete pumps. Pump operators also became more knowledgeable about their operation, and finally concrete producers learned the modifications of standard mix design necessary to produce more pumpable

Figure 8-1 Early concrete pump. (Source: Rexnord.)

concrete. Moreover, the contractor learned to adjust his placing schedule to avoid unnecessary delays that could create line blockages.

TYPES OF PUMPS

Today there are three kinds of concrete pumps:

1. Pumps which force concrete through 4- and 5-in. lines by the positive displacement of the pistons
2. Pumps which displace the concrete by means of compressed air
3. Pumps which squeeze the concrete through a section of rubber hose by means of a wheel rolled along the line

This newer piston-type displacement pump (Figure 8-2), a far cry from earlier pumps, is equipped with a boom that can deliver concrete several stories high effectively.

As in any mechanical operation, the competence of the pump operator and the proper maintenance of the equipment are important. Regardless of the type of pump employed, the best assurance for trouble-free performance is a good pump mix. The drastic modifications in mix designs that used to have to be made with the early pumps are no longer necessary. Still, this does not mean modern pumps will handle just any mix.

Figure 8-2 A top-side pump operator's platform on the Thomsen HP 800 series concrete pump. (Source: Pit and Quarry Publications.)

PUMP MIX DESIGN

The gradation of the fine and coarse aggregate is important. A gap-graded aggregate will often segregate and create line blockage. The proper S/A ratio of the mix must be used for good pumpability. The fine aggregate plays a major role in a pump mix. Experience has shown that if good pumpable concrete is to be produced, the percentage passing the No. 30 sieve should be above 37 percent, and the percentage passing the No. 50 sieve should be 15 to 20 percent.

The normal minimum cement factor of a pump mix is 6.0 bags. The addition of an air-entraining and a water-reducing agent aids in plasticity and reduces segregation. However, with good aggregates and a well-designed mix, concrete of 5½ bags will pump well. In hot weather, concrete with a high cement factor will often cause line blockages when there are long delays in the pumping operation. Under such hot weather pumping conditions, the substitution of a set-retarding admixture for the water-reducing agent is well worth considering.

Crushed coarse aggregate, particularly if flat and elongated particles are present, is more difficult to pump than rounded particles. With the average pump line being 4 in., 1- or ¾-in.-maximum-size aggregate is preferable.

The process of trial and error on a jobsite is a hard way to find out if a concrete pump mix is readily pumpable. When the concrete supplier is new to the pumping game, it would be worthwhile for him to rent a pump and operator for a day at his concrete plant. The pump could be stationed experimentally at several locations in the plant yard, to simulate different pumping jobsite conditions.

Some changes in the mix will usually be necessary. The adjustments should be made carefully; oversanding should be avoided. A deficiency in the coarse aggregate may permit the concrete mortar to flow back through the valves on the suction stroke of the piston and cause segregation. On the other hand, a surplus of fine aggregate will require additional cement to compensate for the increased surface area, increasing cost unnecessarily.

The experimental mixes should have a slump of 5 in. and a cement factor of 6.0 sacks of cement. The mix should contain an air-entraining agent, but an accelerating admixture that contributes to a quicker set should be avoided. Keep making minor modifications to the mix until you have attained the lowest possible pump pressure for pumping ease.

A plastic mix is more important than slump. Study the concrete coming from the discharge end of pump line. Check to see if there are any signs of segregation, or if the mix bleeds excessively. If the pump operator is experienced and knowledgeable, it is advisable to seek his opinion.

When the ideal mix has been attained, the slump, the air content, and the concrete temperature should be taken at both ends of the line. Any differences should be noted and recorded, along with the mix data. It is necessary at this time to mold concrete cylinders of the mix as it enters the pump hopper.

Further, it is suggested that the new mix be redesigned to include a water-reducing agent. Here, one may consider reducing the cement factor by half a sack. Note the change in the pumping performance and in the general appearance of the concrete at both ends of the line. Take cylinders to compare 28-day strengths of the two mixes.

Generally, there will be an improvement in the appearance and performance of the mix with a water-reducing agent (i.e., plasticizer), and the strength should be equal or even higher. However, not all job specifications permit the use of a water-reducing agent, and it is best if the supplier knows the characteristics of his mixes with and without this admixture.

THE CONCRETE SUPPLIER

The concrete supplier, having attained a readily pumpable mix, now must assure himself that the aggregates used in the test remain reasonably consistent. A daily check of aggregate gradation becomes especially necessary now. He cannot wait until the day of a pump operation to look into this phase of his production. It's too late then to compensate for a drastic change in gradation.

Unfortunately, even when good quality control exists, the supplier is not necessarily home free. Not all pumps are perfect, nor are all pump operators sufficiently knowledgeable about the proper maintenance of their pumps. Sometimes, because of imperfections in the individual pump, problems can arise. Sometimes the problem is caused by the laziness, carelessness, or incompetence of the pump operator. Sometimes the problem can develop because of poor, or nonexistent, maintenance of the pump or its lines. Jobsite delays, created by the general contractor, can cause problems.

THE PUMPING CONTRACTOR

The pumping contractor can create unnecessary problems by retaining lengths of rubber hose sections that are no longer needed.

A steel pipe section of the pump line has a lower coefficient of friction than a rubber section. It is for this reason that pumping contractors, especially in hot weather, will use as many steel line sections as possible. Lines get hot from the sun, and additional heat is created by the friction of the concrete moving through the lines. The more resistance to concrete flowing through the pump line, such as bends, elbows, and height, the higher the pressure demand at the pump.

A knowledgeable pump contractor, as shown in Figure 8-3, removes excess line sections as soon as possible. These sections are carried off the concrete placement area and thoroughly cleaned inside and out to eliminate any possible buildup of set concrete within the line.

The concrete pump line moves back and forth with each thrust of the pistons. For this reason wooden cradles are used to support the line and prevent displacement of the reinforcement. Unnecessary excess line sections that must be moved as the concrete placement area moves will often pull the line off the cradles and disturb the reinforcement.

The ready-mix producer may just provide the pump and the operator, but occasionally he may also be the pumping contractor. This is perhaps necessary when there is no other pumping service locally available. This has one advantage in that he knows his pump is well maintained and operable. If there is a local pumping contractor, the concrete producer may have to weigh the benefit of eliminating his pump in favor of the local company, who may be a good customer.

Regardless of who the pumping contractor is, when pumping problems arise the fault is usually placed with the concrete supplier. Nor is

Figure 8-3 A concrete bridge deck being pumped by an experienced contractor who is keeping the pump line as straight as possible and removing excess lengths as necessary. (Source: Pit and Quarry Publications.)

this unreasonable since, in the past, the concrete mix might well have been the cause of the problem. However, as often as not, the problem was created by something other than the mix.

PUMPING: PROBLEM CAUSES

When the pumping operation has been going smoothly and a problem arises suddenly, it may have been created by several factors, which include:

1. One truck may have concrete buildup in the mixing barrel, preventing proper mixing.
2. The delivery truck may not have been given sufficient time to mix the load properly.
3. A slow delivery may have caused a delay in pumping.
4. A long delay caused by more trucks on the job than the contractor could handle efficiently may have created rising concrete temperatures and caused concrete stiffening.

5. A new and inexperienced driver may have fouled up a load.

Only when the concrete producer knows his own house is in order is he mentally free to look elsewhere for the cause of the problem. If his concrete mix has been used successfully in the past, and if his aggregates are the same as previously used, he should look elsewhere for the cause of the problem.

He should see if mortar is leaking from the piston chamber of the pump. A missing or a broken gasket can contribute to decreased pumping efficiency. Mortar could also be leaking from one or more line connection joints. There are metal flanges at each end of a line section that should be covered with a rubber gasket and held in place by a metal clamp. Absence of the gasket, or poorly clamped connection joints, can cause a serious loss of important mortar line lubrication, causing line blockages.

All pump lines have sections of semiflexible rubber hose with a life expectancy of carrying approximately 20,000 cu yd. They are expensive and generally replaced only when necessary. When harsh mixes are being pumped, the life of the rubber section may be reduced to 10,000 cu yd.

With excessive wear these rubber lines will split on the inside, usually at the metal flange at the far end of the section. This line rupture creates a buildup and will reduce the inside diameter of the line, leading to higher pumping pressures and slower production, or will create complete blockage.

Once the problem section is located, experienced pump operators will replace the trouble section. Generally this problem develops at the discharge end of the hose section. The problem section should be replaced; for expediency, reversing the section will sometimes suffice.

All pumps have, or should have, a metal grill on top of the concrete receiving hopper to prevent foreign objects from entering the pump and clogging the line. These grills should have openings ½ in. less in diameter than the narrowest section in the line. This will prevent small chunks of built-up concrete from the concrete truck mixer, or concrete balls, or foreign materials that may be in the aggregates, from entering the lines.

PUMPING LIGHTWEIGHT-AGGREGATE CONCRETE

It has always been advisable to prewet structural lightweight aggregates before use in concrete. When the lightweight concrete is to be pumped, saturating the aggregate before batching is a must. Structural lightweight aggregates, such as expanded shale, have a cellular structure with a high absorption factor. Unless thorough saturation is accomplished, the aggregate particles will absorb mixing water while in the pump mixing hopper and lines, and will stiffen the concrete and cause blockages.

Experience has shown that better pumping performance is obtained by using a natural fine aggregate than by using one produced by lightweight manufacturers. Natural sand is usually delivered to the concrete plant in a premoistened state, and as a result, its absorption factor is nil.

When lightweight coarse aggregate is stockpiled, it should never be stocked in cone fashion. The loads should be dumped in truckload piles close together and kept watered for at least 48 hours by means of one or more lawn-type sprinklers so placed as to cover the entire stockpile area. For the same reasons it is important to consider the moisture content when batching. When in doubt, one can assume a 10 percent average will ensure a safe yield.

A typical mix for 1 cu yd of a 3000-psi lightweight-concrete mix, consisting of lightweight coarse aggregate and natural sand, is as follows:

Ingredient	*Amount*
Cement	6.0 sacks (564 lb), SSD
Sand	1400 lb, SSD
Coarse aggregate	900 lb, SSD
Air	6.0 percent or higher
Water	As necessary to achieve desired slump

Some concrete producers prefer to use a ⅜-in. coarse aggregate, while others feel they have better pumpability with a 1-in. aggregate. Since both work well when properly graded, their comparative cost should be considered when selecting between them. The sand/total aggregate ratio varies with the different manufactured coarse aggregates. It is not unusual to find that some of the more expensive materials cost less per cubic yard of concrete, because of particle structure and weight per cubic foot. If there is a choice in your area, it is suggested that the weights per yard of the different manufacturers' mix designs be compared with cost in mind.

Most job specifications limit lightweight concrete to a maximum weight. Some of these weight restrictions are unrealistic and difficult to attain with the aggregates locally available. Where the weight of the concrete is more important than compressive strength, such as in roof decks, increasing the air content of the mix will reduce weight.

The standard methods for determining air contents are not always reliable for tesing air in lightweight concrete. A roll-a-meter is recommended for better accuracy. The small, handy Chace meter is reasonably accurate in normal air content ranges and slumps. Regardless of the type of concrete being tested, the alcohol used with the Chace meter must be isopropyl alcohol in a minimum 91% solution. Any other alcohol, or the isopropyl type in a solution of less than 91%, will result in erratic results.

PRIMING THE PUMP

The first step in any pumping operation is to prime the pump line with a slushy grout. A mixture of 1 part of cement to 2 parts of sand is usually sufficient to lubricate the line. The grout should be observed carefully as it is being pumped, to check for leaky connections. The concrete should follow the grout as quickly as possible to avoid partial drying of the lubricant.

REMOVING LINE SECTIONS

As the placing of the concrete proceeds, unnecessary sections of the line are removed. Each section removed should reduce friction and pump pressure. Naturally, concrete placed on grade offers less resistance than concrete placed at higher elevations because of the negative head. Concrete pumped at high elevations, and with bends and elbows in the line, is normally more difficult to pump. However, recent pump modifications have made it possible to pump concrete up many floors without problems.

PLANNING THE PUMP JOB

In closing this chapter it should be emphasized that many pumping problems could be avoided by proper planning before the start of the pumping operation. When the placement is of any size or importance, a job meeting should be held with all pertinent personnel present. The method of procedure should be detailed. The number of cubic yards of concrete desired per hour should be explained. Expected delays, if any, due to moving the pump, the length of the delay, and at what yardage the delay will occur should be discussed.

Unexpected delays should be planned for. One man should be given the responsibility to contact the concrete supplier and to hold or slow concrete delivery during these delays. If the ready-mix trucks are radio-dispatched and within range of the concrete plant, contact with the supplier is simple. Otherwise, the plant should be contacted by telephone as quickly as possible.

9 / EXPOSED AGGREGATE CONCRETE

TYPES OF EXPOSURE

One of the many advantages of concrete is its ability to be finished in many different ways to add to the beauty and value of a structure. Stucco was, and still is, a very popular finish and one of the most familiar. Concrete can also be hammered, stippled, ridged, and finished in a great many other ways. Indeed, the number of finishing variations is limited only by the imagination of the designer.

Exposed aggregate finishes, although not new, are gaining in popularity on flat work and on large structures. The desired effect can be accomplished in the plastic concrete, in the semihardened concrete, or in the hardened concrete at later ages.

AGGREGATE SELECTION

Generally, the designer will select the aggregate to be used, specifying its size, texture, and color. In the selection of these aggregates he must be careful to avoid aggregates which contain deleterious materials, especially those containing iron oxides, which will stain the surface.

The durability of the aggregate to be exposed must be above suspicion. The choice of aggregates may be unlimited in some areas, while the selection may be quite limited in others. Cost and availability are decisive factors in the selection. It is also often advisable to specify not only the type of aggregate desired, but also its geographical source. Natural materials of the same type differ in color depending on their geographical location. Quartz, granite, traprock, marble, and gravel come in many different shades. The final selection of the aggregate should not be made until a test panel is studied (2 ft x 2 ft x 2 in.

deep is sufficient), since aggregates will usually appear lighter with a mortar background than in their natural state.

CEMENT

The cement to be used in exposed aggregate finishes should also be chosen carefully. The cured color of the cement paste has a decided effect on the general color-tone quality. Many aggregate producers and ready-mix suppliers have test panels, available for inspection, made with local materials. From these test panels, which are generally left exposed to the elements, the effect of weathering can be determined.

For the experienced mason, producing an exposed surface is not difficult, provided he observes a few simple techniques. Naturally, he cannot expect to place and finish an exposed aggregate surface finish as quickly as he might place and finish standard concrete. He should expect to handle only 25 to 30 percent of the amount of concrete he would normally produce.

HORIZONTAL-SURFACE AGGREGATE EXPOSURE

On a flat surface, when the aggregate to be exposed is top-seeded, a very stiff mix should be avoided, while a slump of 4 in. should be considered a maximum. The concrete should be placed and screeded in the usual manner and leveled and smoothed with a wooden darby. Then the selected aggregate, chosen for exposure, should be cast by shovel or by hand, so that the surface is covered as evenly as possible. The material should be tamped into the concrete surface with a wooden device (a wide board will do), until all the aggregate is embedded. Further embedment and leveling is done with a bull float and then hand-floated until the surface is similar in appearance to a normal concrete slab.

The aggregate exposure operation calls for proper timing. Nothing should be done until the surface can withstand the weight of a man. This time delay is needed to prevent any of the aggregate from being dislodged during its initial exposure when the surface is being lightly brushed with a stiff nylon-bristle broom.

In hot weather, or when large areas of exposed aggregate are being placed, the use of a retarding agent sprayed on the surface during the final floating operation is suggested. This will delay the set of the mortar above the aggregate to be exposed, yet will not interfere with the normal setting time of the base concrete. This action will permit removal of the top mortar more readily and prevent the possibility of aggregate dislodgement.

In the final exposure of the aggregate, the timing is critical and will vary depending on climatic conditions. This final step involves the use of a stiff-bristle brush simultaneously with water. Some masons tie a garden hose to the handle of the broom, adjusting the discharge end to the necessary downward spray, and pointing the hose at the concrete surface in front of the broom. Others employ a helper to saturate the immediate working area.

The exposing operation may have to be done several times, until all mortar residue has been removed. (See Figures 9-1 to 9-4.)

Figure 9-1 Aggregate distribution over fresh concrete. (Source: Portland Cement Association.)

Figure 9-2 Embedding of the aggregate. (Source: Portland Cement Association.)

Figure 9-3 Darbying surface to completely cover aggregate. (Source: Portland Cement Association.)

Figure 9-4 Exposing the aggregate. (Source: Portland Cement Association.)

AGGREGATE SIZE

The exposed aggregate to be used must be thoroughly washed before use for good bonding to the cement paste. An irregular coarse aggregate will bond better than one with a smooth surface. The size of the aggregate should be uniform, whether it is ⅜ in., 3 in., or larger. The ⅜-in. aggregate generally used on exposed aggregate sidewalks will be almost unnoticeable on a vertical wall unless one is up close. Viewed from any distance, the small exposed aggregate on a wall will lose its effect. The same is true of all the aggregate on a building of three stories. Seen from ground level, the upper-storied exposed aggregate often appears to be plain concrete. In general, the taller the building, the larger the exposed aggregate that should be selected.

Suggested Visibility Scale

Aggregate size, in.	*Distance at which texture is visible, ft*
¼–½	20–30
½–1	30–75
1–2	75–125
2–3	125–175

The color of the cement must also be considered, for proper contrast with the color of the aggregate. The only limits in color selection and contrast are those of the designer's own imagination: a variety of colored cements are available, as are colored pigments with ordinary portland, white portland, or other portland cements.

TEST PANELS

Be aware, when looking at test panels, that the color and gradation of the fine aggregate can influence the final color appearance. Therefore be sure that the sand ordered and used on the job is the same as that in the test panel. Different sands will create differences in the cured color of the mortar.

CURING

For best results, the curing process for exposed aggregate concrete should be by ponding or by other methods of moist curing. To eliminate color variations, be certain that the wetting is uniform, with no area left to dry before another. Sprayed liquid curing compounds should never be used, nor should any admixture other than air entrainment be considered.

THE MIX DESIGN

The mix design for exposed aggregate concrete should be different from that for the standard concrete. An increase of fine aggregate is required, with a cement factor not less than 6 sacks. Vertical building panels are usually cast in forms on grade and are not lifted into position until they have attained a minimum strength of 3500 psi.

PRECAST PANELS

The compressive strength of precast panels of exposed concrete is usually 5000 psi or higher. They are frequently constructed in two placements to reduce cost. The unexposed lower section is poured with a conventional concrete, while the upper exposed portion is poured with the selected aggregate and the colored paste desired. Both concretes must have the same cement factor and the same ultimate strength if the same shrinkage factor is to be obtained. The methods of seeding the aggregate vary with the different manufacturers.

A representative sample of exposed aggregate panels is usually displayed by the manufacturers for study. The aggregates, carefully chosen by the manufacturer, are sure to be sound materials with low-absorption qualities.

There are several methods of producing a panel. One method is to place a layer of sand in the bottom of the form and to then place aggregate to be exposed carefully on top of the sand, tamping the aggregate gently to embed it to half its depth in the sand. The surface is then moistened with a fine spray, to settle the sand firmly into place. The matrix desired for exposure between the aggregates is carefully placed on top of this section. Prefabricated reinforcing steel is then lowered into place, and the backup concrete is added to fill the form.

The slump of the backup mix should be about 4 in., and the mix should be vibrated with a screed vibrator. When the panels have cured sufficiently, they are raised and the sand on the exposed surface is washed away. This method, which allows for the use of different-sized aggregates, can be used to achieve bold and variegated exposed finishes. One building with exposed aggregate varying in diameter size from 2 to 6 in. catches the eye of every passerby.

SANDBLASTING AND WATER BLASTING

Sandblasting and water blasting after the concrete is 48 hours old are other methods used extensively. Of late the water-blasting method has gained in popularity partially because of the health hazard created by the sandblasting operation. Further, the aggregate to be exposed can be affected by the action of the hard silica sand particles used in sandblasting, dulling its appearance and muting the color of the matrix.

Whichever method is used, the depth of penetration should never exceed one-third of the thickness of the smallest aggregate particle. It is important that all exposed concrete be blasted uniformly. For this reason, blasting is best left to a professional operator, who knows from experience the amount of pressure required to produce the desired finish and the distance at which his blasting nozzle should be held from the concrete.

FLOOR GRINDING

Floor grinding to expose embedded aggregates can produce a terrazzo-type finish. A coarse grit is used to remove the surface layer of mortar, while finer grits are employed to obtain the desired smoothness. Such grinding will expose air voids which must be filled with each successive operation. The same sand and cement mixture used in the floor concrete must be used to fill these voids to obtain uniformity.

Unusual effects can be produced by varying and mixing the colors of aggregate on walls, beams, and spandrels. The timing of aggregate exposure differs with the method used. Exposure by blasting should be done fairly early, before the concrete has gained too much strength, in order to lower the cost of the operation. Other methods, such as the exposure obtained by the use of mechanical pneumatic tools of various types, should not be attempted until the concrete is at least a month old.

EXPOSED-AGGREGATE BLEMISHES

Exposed-aggregate finishes not only have the advantage of unusual appearance, but also, in the act of aggregate exposure, surface blemishes and discolorations are removed. On walls and columns the stripped surface often shows discoloration between sections because of using form release agents, a change in cement brand, and the use of unclean forms. Further, there are blemishes, such as mortar seepage from between form joints and from the bottom of forms on successive lifts, marring the surface below. Many of these flaws are corrected in the process of exposing the aggregate.

With careful workmanship there are few blemishes in exposed-aggregate construction, but they do occur. However, many masons find it easier to repair blemished areas in exposed aggregate than to patch regular concrete.

METHODS OF REPAIR

The area to be repaired is squared off to a depth of an inch or more, depending on the size of the exposed aggregate, and the repair area is thoroughly saturated with water. A bonding agent of cement and water, mixed to the consistency of thick paint, is applied. The final patching concrete, containing the same matrix used in the original mix, is then compacted into the repair area. When the area to be repaired is small, the aggregate is best pressed into this surface by hand.

Rust stains on vertical surfaces are difficult to remove. Acids or chemicals used successfully on horizontal surfaces, where the stain area can be isolated, are likely to run on vertical surfaces, marring the surface below the stain. Prevention of rust stains is the best cure. Keep steel reinforcement away from the surface to be exposed, and be sure the aggregate does not contain iron oxide.

NO-FINES AGGREGATE EXPOSURE

Finally, there are precast methods of producing wall sections without the labor cost of exposing the aggregate. It is referred to as *no-fines concrete* and is made without the addition of sand to the mix, using clean aggregate ranging in size from ⅜ to 1 in., at the rate of 1 part cement to 8 parts aggregate. Many one- and two-storied homes with this type of surface have been produced in Scotland over the past 30 years. It has been found to be durable and trouble-free.

10/CONCRETE REPAIRS

CONCRETE REPLACEMENT

The complete replacement of faulty concrete should always be a last consideration. Although there are still occasions where removal is necessary, such as an old, badly broken sidewalk, for example, most concrete structures are rarely beyond repair. Because of the more recent introduction of specialized repair products and their methods of application, problem concrete can be restored and made stronger and more durable than the original.

REPAIR PRODUCTS

The unlimited selection of repair products available makes choosing the right one for the individual problem important. These products are sold in a wide range of prices, and it is not always necessary to purchase the more costly product when a less expensive one will solve the problem.

How long a repaired area will remain serviceable will depend on the exposure to which it will be subjected. Also the location of the problem will determine, to a large degree, the product for best results. One material used to repair a honeycombed area on a vertical surface may be a poor choice for a scaled floor. It is for this reason that there is such a wide selection of manufactured materials formulated for special purposes.

CRACKS

The repair of concrete cracks is probably the most common. They are caused by a variety of factors. There are two classes of cracks: active and dormant. A dormant crack, such as a drying shrinkage crack, is one not likely to change in character, and may well be ignored. Cracks that increase in length and width or show movement under loads are considered active. Such cracks should be repaired as soon as they are noticed, to prevent their development of a serious problem at a later age.

CRACKS: CAUSES

Unfortunately, cracks are caused by so many different conditions that the reasons for their appearance are not readily determined. Some are created while the concrete is in the plastic state or during the initial set, as a result of one or more of the following:

1. Settlement due to unstable subgrade
2. Subbase paper rupture
3. Poor form construction
4. Lack of, insufficient, or improper placement of reinforcement
5. Rust on reinforcement steel
6. High-slump concrete
7. Improper or insufficient vibration
8. Lack of curing
9. Volume change due to settlement of solids in plastic concrete
10. Heavy ground vibration nearby, such as from a pile driver
11. Stripping forms before the concrete has cured sufficiently
12. Lack of or insufficient expansion and control joints

The decision on whether or not to repair some of these cracks will depend, to some extent, on the subsequent use of the faulty concrete area. A dormant crack on a concrete floor which is to be covered by tile or a carpet requires little or no attention. This same type of crack appearing outdoors should be repaired to prevent its becoming active under adverse weather conditions. Indoor surface cracks are less likely to become more serious under normal conditions than would outdoor cracks that are subjected to greater changes in volume.

STRESS CRACKS

Although some cracks develop during or shortly after concrete placement, others appear weeks, or months, after the concrete has hardened.

Frequently, these cracks are due to loads applied before the concrete has gained sufficient strength to sustain the imposed stress.

When 3500-psi concrete is placed, it is not expected to reach that compressive strength until a month later under normal conditions. In cold weather, which is unfavorable to proper curing, it may be 6 weeks or longer before that strength is reached. Concrete has low flexural strength, seldom above 650 psi, which is related, to some degree, to the compressive strength. If both the above factors are kept in mind, it is easy to see that a load on a concrete floor before it has reached its full required strength may well cause cracking.

After the initial volume change in plastic concrete, which results from some settlement of solids, and the loss of some of the mixing water due to evaporation, concrete will continue to change in volume in the hardened state. Changes in temperature, freezing and thawing, exposure, and wetting/drying cycles create concrete volume change.

The lack of or improper placement of control and expansion joints where volume changes are *excessive* will create unnecessary cracks. Concrete placed on a frozen subgrade, nearby blasts, improper vibration, and other influences in combination cause problems in hardened concrete.

EPOXIES

When the concrete crack presents a serious structural defect, the cost of the repair product should not be considered if the material chosen can permanently cure the problem. Here, one might consider epoxies. One epoxy product, specially manufactured for such a problem, has worked very successfully over the years. Concrete cores taken from a repaired section of the crack had strengths higher than cores taken from a nearby sound section of the same area.

The particular epoxy mentioned here is a two-component system, which produces a 100 percent solid, modified epoxy resin which is injected into the crack and is self-bonding. True, the process is more expensive than some others, not only because of the cost of the material but also because of the method of application. Special entry ports for the injection of the material are installed, where required, along the line of crack. The port-free areas are sealed on the surface of the crack with a special hydrostatic cement paste to prevent leakage. After the crack has been prepared, the resin is injected into the ports, under pressure, until the entire crack has been filled. Later, the ports are removed, and the crack area is ground to an even surface.

A big advantage in the epoxy systems, as with magnesium phosphate cements, is the ability of these repair products to be applied in temperatures below freezing. Some epoxies permit their application to moist, wet, and dry surfaces. They harden rapidly and have been used successfully underwater in the repair of dams. However, the use of most epoxy products requires the employment of knowledgeable contractors familiar with the various products, the proper preparation of the problem surface, and the correct method of application.

PREPARATION OF REPAIR AREA

Regardless of the repair product selected, the success of the project will depend not only on the material used but also on the preparation of the area to be repaired. Both improper surface preparation and failure to follow the manufacturer's recommendations for the use of the product are sure to guarantee a short life of the cure.

Recommended procedures for proper surface preparation are listed:

1. All loose material in small areas, such as chuckholes, should be removed with the use of pneumatic hand tools. Broom and vacuum the hole for the bonding agent to follow.
2. On large areas, such as worn or scaled surfaces, the old surface can be cleaned by sandblasting or water blasting, by using a jackhammer, by being scarified, or by scabbling.
3. All dust and debris must be removed to assure good bond of the repair material. The use of brooms, water blasting, and/or air blasting will usually produce the desired result. When brooms are used, however, it is good additional insurance to vacuum the area for the removal of very fine dust particles.
4. The problem area must be cleaned of any old existing coatings. Remove all oils, greases, dirt, and wax solutions. If the old surface is only slightly soiled, the use of chemical surface-cleaning agents may be sufficient. These surface cleaners should be washed off thoroughly after use with soap and water, then hosed down.
5. Paint deposits or bituminous materials are best removed by flame-treating.

Other methods for removing stains and blemishes will be found in Chapter 11.

SLIPPERY CONCRETE

Many concrete floors without serious defects become slippery from foot traffic and require nothing more than a nonskid surface paint. A reliable nonskid paint is reasonably inexpensive and will stand up well. Where the surface to be treated will receive the kind of wear created by forklifts, on automobile ramps, or on chemical-manufacturing plant floors, epoxy coatings may be considered. They are very durable, resist heavy traffic wear, and can be applied in smooth or nonskid formulas. However, regardless of the repair product selected, the old surface must be thoroughly cleaned before application.

The problem of a slippery floor is obvious, as is the cause. In many cases we are aware of a problem but do not know the cause. Before any decision can be made regarding the repairs, the product to be used to repair the problem and the cause of the problem must be determined. Sometimes the cause of deteriorated concrete is readily apparent, but frequently the reason for the deterioration is unknown, and a contractor should be called in to determine the real cause and advise on the cure.

ALKALI EXPANSION DETERIORATION

Some problems of concrete deterioration, such as alkali expansion, may not appear for several years. Depending upon the depth of the erosion, such deteriorated concrete may have to be replaced. Alkali expansion is activated in the presence of moisture, and if the erosion within the concrete is not too severe, the application of a waterproofing material may stop further erosion. This may *eliminate* the problem, or at least *delay* further deterioration. This questionable cure is often worth the gamble, as it is reasonably inexpensive.

Along with the newer repair products on the market, some of the older products are still reliable and fill a need. Many have been improved, are inexpensive, and can be used by the average handy man. A visit to a local hardware store or lumberyard for help is usually sufficient.

REPAIR PRODUCT CAUTION

Many repair products, such as epoxies, should not be used by a do-it-yourselfer, or by anyone not familiar with the possible dangers in their use. A few people are allergic to the fumes of epoxy resins and react badly. Depending on the temperature and humidity, the product pot life varies, with the average being 30 minutes before the material hardens and becomes useless. A special solvent is required to keep tools free of epoxy buildup. There are other factors involved, but those mentioned above should be enough reason to leave epoxy applications to the knowledgeable.

Most epoxies are manufactured in a two-component system. Each component alone has a long shelf life, but when both are added together and mixed, the time span for application is very limited. The mixture hardens quickly and gains amazing strength in a few hours. For this reason it is an excellent repair product for concrete areas that must be opened to traffic as quickly as possible.

Epoxy resins are organic compounds that develop excellent strength and adhesive properties. They have good physical stability and are resistant to many chemicals. They are crack-resistant and resist water penetration. They can be applied to moist or wet surfaces in temperatures above freezing and applied to dry surfaces at 0°F.

The average epoxy manufacturer produces a full line of formulas to cure most concrete problems or to preserve concrete from future attacks of destructive forces.

A few of the epoxy products are listed below. They are, for the most part, modified epoxy or synthetic resin systems.

1. Penetrating sealers
2. Waterproofing systems
3. Concrete- and steel-bonding systems
4. Anticorrosive systems
5. Concrete-injection systems
6. Mortar-patch systems
7. Antiskid-surface systems
8. Paints and coating systems
9. Sealers
10. Primers

In item 6, silica sand is added to the two resin components after they are mixed. When the repair hole is deep, special coarse aggregate may be added for bulk. A very fine aggregate, similar to blasting sand, is introduced to the item 7 product resins to produce a durable antiskid surface. These two products are considered three-component systems, while all the others are two-component systems.

LATEX- AND ACRYLIC-MODIFIED CEMENT MORTARS

The more conventional patching materials, such as latex- and acrylic-modified cement mortars, with the addition of a synthetic latex, are excellent products that perform well, are easy to apply, and are less expensive than epoxy materials. These products produce a greater internal strength than do plain cement mortars. They are more flexible and durable, and will resist freeze/thaw cycles better than the conventional patch. Although they are not as resistant to chemical attack as epoxy-based materials, they are a popular choice for areas not normally subjected to such deteriorating forces.

Unfortunately, many repair materials do not match the color of the surrounding concrete. If a mortar or concrete patch is used, with the same cement as in the original mix, there probably will be a difference in color, but the patched area will match the unpatched area in time. There are, however, many problem repairs where slight differences in color are of little concern. On highways or bridges, for example, the difference in color in the concrete pavement will blend in reasonably well under car and truck traffic. It is more important to select the most durable product needed than be concerned with the difference in color in some areas.

GRINDING AND MUD JACKING

It is not always necessary to repair concrete surface problems with materials. When the problem of slab settlement exists, several other solutions are possible. If the settlement is slight, grinding the higher slab to an equal level with the lower section is quick and inexpensive. Where the difference in levels is serious, mud jacking should be considered. Holes are drilled into the settled section of the slab, and a slurry of grout or similar material is pumped into the holes to lift the problem slab to its original position.

When the settlement of a slab amounts to nothing more than a trip step or a jolt spot to forklifts and other wheel equipment, an epoxy mortar is an acceptable cure. Many repair products do not lend themselves to featheredging, and they will break down at the thinnest edge after a short time. Once they start to wear at the edge, further destruction of the patch is inevitable.

FREQUENT PROBLEMS

Following headings cover the more frequent problems requiring repairs.

Popouts

Popouts (Figure 10-1) are caused by unsound coarse aggregate particles near the surface of the concrete. These particles readily absorb water, and under freezing conditions, they expand and destruct, popping off the thin surface layer of mortar above. The resultant hole is a perfect spot to hold water which can freeze in cold weather and cause further destruction.

Further popout problems are best avoided by treating the entire surface area with a water-repellent liquid sealer to prevent further water absorption. The holes may be filled with acrylic compounds or latex-modified mortars that are self-bonding.

Figure 10-1 Typical popout. (Source: Portland Cement Association.)

Dusting

Dusting is indicated by the presence of a fine whitish powder on the surface of floors. Usually, dusting is caused when the surface is troweled while bleed water is still on the concrete surface. One or two applications of a solution of magnesium fluosilicate and zinc fluosilicate will usually cure the problem. This solution is sold in most hardware stores and lumberyards under various trade names, in both liquid and powder forms.

Saturate the dusted surface by sloshing the solution on and spreading it over the entire area with a mop or broom. Bubbling will be noticed during the application and is to be expected. A definite improvement will be noted within 24 hours. Where necessary, make a second application in the same manner.

Honeycombing

Honeycombing, as shown in Figure 10-2, can be caused by using a cement mix with excessive coarse aggregate, by leakage of mortar through the forms, by very low slump concrete, and by improper or insufficient vibration. Besides creating an unattractive appearance, the honeycombed area can be a weak link in the structure.

When preparing small honeycombed areas, as when repairing small holes, people are likely to scant on surface preparation. The result of this frequently causes the patch to dislodge. Removal of all loose aggre-

Figure 10-2 Honeycombing. (Source: Portland Cement Association.)

gate particles is essential for satisfactory repair. All exposed aggregates and crevices must be wetted and a good bonding agent applied. The more commonly used materials are epoxies, expanding mortars, and portland cement mortars. The material chosen should be dense and must be pressed firmly into place.

When the area to be repaired is large, it might be better to use a form. Some are formed completely, and the repair material is injected under pressure through ports in the form. Another method is to leave an open space at the top of the form, and then to place concrete, similar to the original mix, through the opening and compact it well. After the concrete has hardened enough to permit form removal, the open area previously left unrepaired is then patched.

Cavitation

Cavitation, very similar in appearance to honeycombing, can appear months or years after the concrete has been placed. Wherever there is a flow of water passing a protruding object, or an uneven vertical wall, cavitation may appear on the downstream section of the irregularity. It has been suggested that the bubbles created by the protrusion burst, and so create the phenomenon. Use the same repair techniques for cavitation as for honeycombing, being sure to produce an even surface.

Cold Joints

Cold joints in mass concrete, as shown in Figure 10-3, are caused by improper placement of the plastic concrete. Delays in placement, failure

Figure 10-3 Cold joints.

to place the concrete in horizontal lifts, and inadequate vibration are among the many reasons for cold joints. True, the strength of concrete is dependent on the compressive strength of the concrete mix and the amount and placement of the reinforcing steel. Yet in mass concrete, the steel does the major job with the concrete being, more or less, a filler. An overly rich mix can only add to the heat of hydration and help create a cold joint situation. Regardless of the ambient temperature, the addition of a set-retarding admixture will reduce the potential for cold joints.

When a cold joint occurs during placement because of delivery delays, and the concrete cannot be revibrated to a plastic state, a 1-in. layer of grout, before continuation of placement, will pay dividends. The epoxy injection system is the best method of repair.

Water in Basements

Water in the basement is a frequent problem in residential construction. Although a poured concrete foundation is acknowledged to be superior in strength and density to hollow core block, it is not a guarantee of a dry basement.

Ideally, prevention of water seepage into a basement is preventive insurance in the construction of the foundation. A footing should be keyed, cleaned, and bonded *before* the foundation wall is erected. A drainage sleeve should be installed in the footing to permit water to escape from under the basement floor to an outside source, either to a dry well or to a natural-site slope. Additional sleeves should be installed through the wall to provide for the subsequent entry of utilities.

A recent survey of home builders regarding damp basements indicated homeowner complaints were about equal in both poured concrete foundations and block construction. Many who had previously treated the outside of the foundation wall before backfilling with waterproofing materials stated that a light coating of a bituminous material on the wall, followed by an application of 4-mil polyethylene sheeting over the coating while it was still tacky, had drastically reduced complaints. This sheeting, when well-overlapped at the joints and sealed, had provided the best results.

This new system alone will, naturally, not cure all problems of basement water seepage. In construction where there is poor soil drainage, it is evident that drains should be provided around the outside perimeter of the building. At least 4 in. of coarse aggregate should be placed below the basement floor to reduce hydrostatic water pressure. Before the concrete basement floor is placed, a concrete bonder should be applied to the footing concrete and the wall to which the floor concrete will abut.

However, in spite of precautions, or in homes without these innovations, there may be water problems. In structures with poured concrete walls, the tie-rod holes should be patched with hydraulic cement to prevent seepage. This same product can be used to fill floor cracks and to stop seepage through the floor at the inside perimeter caused by concrete shrinkage. If these openings are too narrow for good application of the cement, or other repair materials, they should be widened to a depth of at least ½ in.

Any hole in the inside vertical wall should be patched with hydraulic cement and the entire wall area treated with a reliable waterproofing material. However, keep in mind that in most areas the building contractor is responsible for any defect in construction for a period of at least 1 year. Regardless of the required responsibility of the builder, most reliable contractors will respond to a complaint call on any construction defect within a reasonable time.

Occasionally, water in the basement is caused by rainwater from a downspout. An elbow at the bottom connected to an extension drain away from the foundation will usually cure the problem. If all else fails, the installation of a sump pump may be necessary.

Efflorescence

Efflorescence is a white stain on concrete surfaces caused by the penetration of moisture through the concrete. These stains are easily washed off with soap and water followed by rinsing. Should the stains persist or be severe, applying dilute muriatic acid, and then rinsing it off well, will generally eliminate the problem.

REPAIR OF PITTED OR SCALED SURFACES

Repair of pitted or scaled surfaces will depend, to a great extent, on the traffic the new surface will receive. Whatever the repair product decided upon, the problem surface must be carefully prepared. The pitted surface must be removed to a sound substrata by sandblasting or water blasting; by machine tools, such as the scabbler; or by any other method that will remove the faulty concrete. Grinding should be avoided, as it leaves the cleaned area too smooth. A rough surface will provide better bond to the repair material.

It is common practice to etch the area with a 10% solution of muriatic acid for the final cleaning, but this should not be attempted by a novice. The acid can burn the skin and change the pH of the old concrete, thereby reducing bond. When the acid method is used, be sure to water-flush the surface thoroughly to remove all traces of the solution.

For small repair work on pitted or scaled areas, many contractors favor the use of acrylic-based materials. They can be purchased in bags, in dry form, and after being mixed with water they can either be applied with a broom, for a nonskid surface, or be troweled. They have the advantage of being self-bonding, are easy to apply, and are resistant to deicing chemicals and salts.

Whether the repair area is large or small, the perimeter should be squared for best results. Whenever possible, try not to leave a prepared area that would need a feathered edge.

CHOICE OF REPAIR MATERIALS

The choice of the repair material to be used will depend not only on the particular nature of the problem but also on the function of the structure, the availability of equipment and skilled manpower, the relative importance of appearance, and of course the funds available for the repair.

Listed alphabetically below are some of the more frequently used repair materials.

Acrylic-modified mortars

Elastic sealants

Epoxies

Expanding mortars

Flexible epoxy resins

Latex-modified compounds

Portland cement concrete, grout, or mortar

Quick-setting materials

For active cracks, for replacement of expansion joints, or for use between any two surfaces that are subject to movement, elastic sealants are a good choice. They have the ability to return readily to their original shape after deformation. Many of these materials maintain a bond between two surfaces even when subjected to great stretching or torsion forces.

Expanding Mortars

Expanding mortars, grouts, and concretes have been developed to eliminate product shrinkage. There are several different manufacturing processes used in these types, and care should be given to the selection of the right kind.

Quick-Setting Materials

Quick-setting materials, usually referred to as *water plugs, wet plugs,* or *hydraulic cements,* set rapidly and harden in a matter of minutes. These materials are useful where water is coming from a hole under pressure.

Latex-modified concrete has become very popular because of its adhesive properties and high compressive and tensile strength. It is reasonably flexible, has a low water-absorption factor, and is durable. It is recommended for the repair of concrete roads, bridges, and heavy-duty floors.

Portland cement concrete has several advantages as a repair material for deep repairs. It has the same thermal qualities as the concrete to which it is being bonded. It is readily available, well known and understood, and is reasonably low in cost.

Portland Cement Grout

Portland cement grout is generally used only where the opening to be repaired is small, and where shrinkage can be tolerated. For best penetration, it is usually pumped under pressure into these openings. However, it is preferable to use an epoxy injection resin if the repair to be made is important.

Portland Cement Mortar

Portland cement mortar is used for a variety of repairs. It should never be used on very shallow problems, as it will usually not stand up well. All mortar repairs should be at least 1½ in. deep and never featheredged. The sides of the repair area should be vertical to the full depth of the patch.

As with any concrete repair, when portland cement is to be applied, the repair area should be saturated with water in order to prevent absorption of water from the fresh patch. A bonding agent should then be applied, and while the bonding agent is still tacky, the stiff mortar should be carefully compacted into place. As soon as possible place wet burlap over the patch and keep it moist for at least 3 days.

Epoxy products have been thoroughly discussed earlier in the book in detail. It should be mentioned here, however, that the epoxies of a few years ago have been vastly improved and should be seriously considered where the repair to be made is in an important part of a structure.

CONCRETE BRIDGES: DETERIORATION

There have been many articles in the past few years on the deterioration of bridges in the United States. It is estimated that well over 25 billion dollars will be needed to repair or replace thousands of bridges across the nation. Improper maintenance and the lack of bridge inspection are considered the major causes.

It can be readily understood why some bridges, constructed before the advent of air-entrained concrete and other technological advances were introduced, are in need of repair. However, bridges built in the nineteen-seventies that show evidence of badly spalled surfaces are more difficult to understand.

Studies determined that the major cause of surface deterioration was the rusting and placement of the reinforcing steel embedded in the concrete. Reinforcement placed in the concrete with initial evidence of rust-scaling was a contributing cause. *Reinforcement placed too close to the surface was a strong factor.* Water and deicing chemicals, entering the concrete, eroded the rebars and contributed to the failure of the concrete to sustain loads.

Recently, newly made rebars have been coated before delivery to the jobsite with lifetime protective coatings that will assure improved performance. Bridge designers are placing the steel bars a minimum of 3 in. below the surface of the concrete with far better results.

SYNTHETIC LATEXES

Synthetic latexes are plastic particles (polymers) dispersed in water. The resultant fluid, which is milky white, is manufactured with various percentages of solids. For use on deep repairs these fluids are added to a standard concrete mix to produce a mortar with superior durability qualities. The coarse aggregate used in the concrete is generally a ⅜-in. size. For thin patches synthetic latexes are added to the customary cement/sand/water mortars.

LATEX-MODIFIED CONCRETE

Latex-modified concrete has excellent bonding characteristics. It resists alkalies and dilute acids, has low water absorption, good freeze/thaw stability, and featheredges well. Like many of today's repair products, it will be even better in the future. Modern technology and continued research and development will continue to improve older products and introduce new and better ones.

11 / Removing Concrete Stains

The removal of surface stains can be a simple process or a complex one depending on the nature of the stain. Except for the detraction from the appearance of the surface, many stains are harmless to the concrete and will remain so. There are a few (such as acid stains, for example) that, if left unremoved, will create concrete deterioration. Stains may occur in concrete in the plastic state, such as absorption of form oils, or they may be created by spillage of chemicals or other staining agents months or years after the concrete has hardened.

STAINS VS. DISCOLORATION

Some of the problems are concrete discolorations rather than stains. One area of the concrete may differ from the rest, possibly because of excess water on the subgrade at that spot. The concrete placed above that water would then appear lighter in color. There is the possibility that one load of concrete was a different color from another because a different brand of cement was used. A change in the source of the fine aggregate in the mix may cause a color variation.

Concrete containing calcium chloride will usually dry darker than concrete without this often-used accelerator. A drastic change in slump between loads will vary color. Overtroweling or coating dry cement on the concrete to absorb bleed water will produce burn marks or discoloration. Determination of the cause is necessary before prescribing a cure. Moreover, it is often best to do nothing if the concrete is less than a few months old. When some discolorations are exposed to sun and weathering over a period of time, the problem area may blend in nicely with the rest of the surface.

TREATMENT FOR DISCOLORATION

A first, generally reliable treatment for discolorations is the application of phosphoric acid, which can be purchased at any local chemical supply house. An acceptable solution for discolored concrete is 1 pt of the acid diluted with 2 gal of water. Mix well and brush over the entire surface. It may bleach the concrete to a whitish color, but the concrete will gradually return to normal over a period of time. If the variation in color is slight, one might try a similar solution of laundry bleach, which is sometimes effective.

CONCRETE STAINS

Concrete stains are more troublesome than discolorations. The cause of the stain must be determined before successful treatment can be undertaken. Coffee stains, tar, paint, grease, ink, and the many other types of stains all require different methods of removal. Some stains may require trial and error before a cure is obtained.

Stains can be removed not only by chemical treatment but also by the mechanical methods of grinding, sandblasting or water blasting, scarifying, or steam-cleaning. Regardless of the method selected, it is necessary to protect nearby objects susceptible to damage from the treatment—wood, siding, glass, metal, machinery, heaters, and air-conditioning units. It should also be kept in mind that hot-air heaters and air conditioners can bring dust and chemical odors to other parts of the building during the stain-removing process.

Some stains, such as chewing gum or tar, for example, generally remain on the surface with little or no penetration into the concrete itself. Liquid stains do penetrate into the concrete, with their depth of penetration depending largely on the porosity of the concrete and the type of finish the concrete had initially received. A hard troweled concrete finish will normally not absorb stains as will a wood-floated or broom finish.

ACID AND NONACID CHEMICAL TREATMENTS

Most nonacid chemical treatments, when carefully applied, will not injure the concrete. If acid chemicals are used, the surface to be treated should first be water-saturated to dilute the acid absorbed by the concrete. It is best to experiment on a small test patch to determine the success of any chemical treatment before attempting large areas. Applications of this kind should be left on the test patch for short periods only to determine effectiveness. Two light applications are often more successful and less harmful to the concrete than one heavy dose.

The manufacturer's instructions for use of any chemical should, of course, be carefully followed. When such chemicals are used indoors, proper ventilation is necessary. Avoid any contact of the chemical on the skin, wear goggles, and be careful not to inhale the fumes.

POULTICES FOR NONPENETRATING STAINS

Many concrete stains can be removed without mechanical or chemical treatment. A poultice can be applied to the nonpenetrating stains, such as those from caulking compounds and chewing gum, with little effort and good results. Ice or a poultice (i.e., a mixture of flour and denatured alcohol) can be applied to this type of substance to harden it and make it brittle for easier removal. After this treatment the spot should be washed with hot water and a scouring compound, applied with a stiff-bristle brush, to remove any remaining residue. Finally the spot should be rinsed well with water after the treatment has been completed.

A poultice should be applied as a smooth, nonflowing paste and troweled over the problem area in a thickness of ¼ in. or slightly more. The stain-removing product used in the paste dissolves the staining substance, which is then absorbed into the poultice. After the poultice has dried, it is easily removed. The poultice method has the advantage of preventing the stain from spreading during treatment.

EPOXY, GREASE, AND OIL STAINS

Epoxy stains are difficult to remove by chemical treatment and may best be eliminated by sandblasting. In the case of grease or oil stains, sandblasting should not be considered, as it may well drive these highly penetrating materials deeper into a porous concrete.

If the cause of a stain is difficult to determine, removal might first be attempted by brushing the area with a strong detergent or scouring powder. A bleach, such as Clorox, may well be worth testing on the stain. In any event, scrape off as much of the stain as possible before any application.

RUST AND PAINT STAINS

Rust stains, unless deeply embedded, may be treated by mopping the area with a solution of 1 lb of oxalic acid per 1 gal of water.

Wet paint stains will only spread further if they are wiped up. Instead, the paint should be absorbed with paper towels and the residue scrubbed with scouring powder and water. Old, dry paint stains should be scraped off as much as possible, and a poultice, saturated with a commercial paint remover, should be applied.

ASPHALT, TARS, AND PITCHES

Asphalts, tars, and pitches have a good adhesion to concrete and are difficult to remove. Before applying any treatment, scrape off any excess bitumen and scrub the surface with scouring powder and water. CAUTION: On concrete surfaces, do not use steel brushes, since particles of the metal may break off and later create rust stains. Do not use solvents, since they will cause the stain to penetrate deeper into the surface.

One method, sometimes successfully used in removing such materials, is to apply a bandage saturated with a solution of equal parts of dimethyl sulfoxide and water. Let stand for 1 hour, then scrub the treated area with a stiff brush. (A bandage is a few layers of white cloth.)

MOSS

Moss occurs frequently on concrete surfaces that are in a constantly damp and shaded location. Ammonium sulfamate, available from most garden supply stores, has been used successfully in removing such growths. Should a powdery deposit be left on the surface, it can be removed by washing with water.

LUBRICATING AND PETROLEUM OILS

Lubricating oil and petroleum oil readily penetrate into the concrete surface. If the spillage is noted, it should be soaked up immediately with paper towels or absorbent cloth. Repeat, *soak, don't wipe,* as wiping spreads the stain. Cover the spot with a dry powder, such as flour, dry cement, or similar absorbent material, and leave it for a day. Repeat the application as frequently as necessary.

PATIOS

Backyard patios frequently used for eating and entertaining are subjected to many different types of stains. As one treatment may not cure them all, it may be easier to cover the entire area with an acrylic or latex emulsion compound. They are readily available, reasonably inexpensive, easily applied, and durable. A thin coating is sufficient to cover most stains; or one may consider covering the patio, or concrete porch, with outdoor carpeting.

12 / Inspection and Testing

COMMERCIAL TESTING LABORATORIES

A commercial testing laboratory provides an important and necessary service to the construction industry. Although a large portion of the work in a local testing lab is confined to concrete testing and inspection, the lab's personnel are generally equipped to conduct tests on most materials used in construction.

A testing laboratory is rarely a highly profitable business. Testing equipment is expensive, and key personnel must be paid on rainy days and over the winter, when construction is at a standstill. Trained technicians are rarely available for hire and must be trained, often for long periods of time, before they become skilled enough to do their assigned tasks without supervision.

FIELD AND LABORATORY PERSONNEL

Testing laboratories must have sufficient field personnel to provide inspection at the height of the busy season and yet keep them on the payroll when things are slow. Frequently the laboratory, unable to afford this expense, must lay off inspectors and then subsequently hire new personnel who must be trained.

INSPECTION

Inspection is good insurance for the architect, the structural engineer, the contractor, the concrete supplier, and the eventual owner of the project. No matter how well designed and properly specified a project may be, it can lose its value with inadequate inspection. Unfortunately, however, there are occasions when inadequately trained employees are sent to inspect important projects. The cost of such inspectors is wasted, and the money is better spent on retaining the more reputable though higher-priced testing company.

TRAINING AND EXPERIENCE

Along with training and experience, tact and judgment are equally important requisites for a good inspector. He should know enough to remain simply an observer when the contractor is doing his work properly. Likewise he should be able to appreciate the circumstances under which the contractor is operating and therefore ignore minor infractions of the specifications which are not likely to be harmful to the project. Primarily, the inspector should be interested in results rather than methods.

An engineering education is not a necessary qualification for an inspector. There are three types of inspectors: the practical type who has worked some time on construction, the technical one who has received training in a technical school, and the ideal one who has a background of both. Generally, on a large project, there are several inspectors for various phases of the work, headed by an experienced chief inspector who is responsible for the performance of all his men.

TESTING LAB: RESPONSIBILITY

The testing laboratory is responsible for the following:

1. Designing concrete mixes specified for use on the project
2. Sampling, identifying, examining, and testing concrete materials
3. Inspecting the concrete plant's daily batching operation and checking the gradation of aggregates for conformity
4. Testing the slump of the concrete, testing the air content, and preparing the concrete specimens for laboratory strength tests
5. Preparing records and reports

INSPECTION: SCOPE

Along with performing these standard duties, an inspector may be called upon to observe the mixing, conveying, placing, compacting, finishing, and curing of concrete. Generally, the testing lab is retained for the testing of the soil, along with concrete inspection. Usually, along with the laboratory inspector, a representative from the architect's or structural engineer's office will check the forms and the reinforcing steel before the concrete placement. He is usually present during the concreting operation and will remain until the concrete is finished and cured.

When a laboratory is retained for a project, one of the first duties of the laboratory personnel is to visit the plant of the concrete supplier to obtain samples of the cement, aggregates, and admixtures required

Compressive Strengths
(psi at 28 days)

	Trial mixes		
	No. 4	No. 5	No. 6
Cylinder H	5535	5022	4333
Cylinder I	5570	5128	4456
Cylinder J	5659	5181	4421
Average, psi	5588	5111	4403
W/C Ratio	4.45	4.60	4.90

Water/Cement Ratio Curve

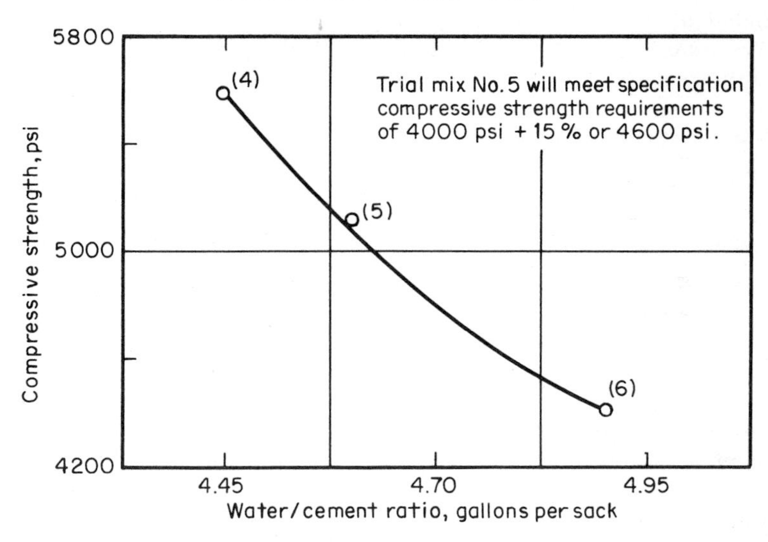

Figure 12-1*a* Water/cement ratio curve with 28-day test results reported. (Source: Fairway Testing Laboratory.)

in the mix design. These materials are then taken to the laboratory, and the aggregates are tested for gradation, specific gravity, absorption, the presence of deleterious materials, and, when necessary, soundness.

If the materials meet specifications, a series of trial batches, usually of 1 cu ft, are conducted. These trial batches are conducted to establish the S/A ratio for workability and finishability with the specified slump. Several batches are then made with several cement factors.

Five cylinders are molded from each of these mixes: two are tested for compressive strength at 7 days, and the other three are tested at 28 days of age (Figure 12-1*a* and *b*). A W/C ratio curve is charted (as

Class BAMix

	Trial mixes		
	(M4)	*(M5)*	*(M6)*
Cement, lb	658	611	564
Sand (SSD), lb	1230	1290	1340
Stone (SSD), lb	1820	1820	1820
Pozz. 122R, oz	24.5	22.8	21.0
MBVR, oz	5.2	4.9	4.5
Total water, gal	32.5	32	31.5
Cement, bags/yd	7.0	6.5	6.0
Slump, in.	5	5	5
W/C ratio	4.65	4.90	5.25
Air content, %	6.0	5.5	6.0
7-day compressive strengths, psi			
Cylinder F	3979	3413	3272
Cylinder G	4032	3466	3325
Average	4005	3439	3298

Figure 12-1*b* Concrete mix designs used with 7-day test results reported. (Source: Fairway Testing Laboratory.)

shown in Figure 12-1*a*), and the mix, or mixes, with the strength and W/C ratio required, plus 15 percent, is selected. The additional 15 percent is added on the assumption that the concrete cylinders made on the jobsite will not be made under the same ideal control conditions as those made in the laboratory.

Cylinders are stripped of their molds 24 hours after they have been cast in the laboratory, and they are placed in a curing room under controlled temperatures of 70°F ± 5° and kept moist at all times until the day of test. They are then capped and tested for compressive strength. The results are recorded in pounds per square inch (psi).

Commercial testing laboratories can be retained by the hour or by the day for the testing and/or inspection of concrete. Should the contractor wish to cast his own cylinders, the laboratory can be hired, at minimal cost, to pick up these cylinders from a jobsite; transport them; cure, cap, and test them; and report the test results.

They can be hired as troubleshooters to help solve a concrete problem—by the engineer, by the contractor, by the ready-mix producer, or by a homeowner. They can also be retained on a daily or monthly basis by a concrete producer to establish and maintain quality control. Under this arrangement a lab technician would appear periodically to check aggregate gradation, admixture dispenser output, and scale accuracy, and to follow concrete delivery trucks to take slump and air tests before cylinders are cast. A report of the technician's daily activities would be sent to the producer, and, later, a report on the compressive strength of the cylinders made.

A commercial testing laboratory is ready to serve anyone interested in receiving or producing quality concrete. When a problem arises, the lab must be honest in appraising the fault and report the results of its findings accurately, whether or not such findings may displease the client. If a man wants to be popular and well-liked, the inspector's job is probably not for him. Popularity is not his goal; respect is.

INSPECTOR'S AUTHORITY

Most inspectors must know their job and know what they are supposed to do, but the degree of authority they should enjoy in enforcing regulations is open to question. Certainly they should have the right to prohibit the placing of concrete that obviously does not meet specifications. Frequently, however, once the job is underway, the rejection of a load of concrete with excessive slump is overruled by the contractor and the concrete is used. This situation is more likely to occur when the inspection agency is retained by the contractor rather than by the architect or engineer.

It is generally agreed that an inspector's authority should be limited. The rejection of a load of concrete for a *minor* infraction of the specifications, when there is no other truck on the job, is stupid. Such a rejection may well create a cold joint, which is far more serious than the placing of *slightly* wetter concrete than specified.

REJECTION OF CONCRETE

Probably the most troublesome individual on a construction site is the overzealous or self-important inspector, the kind who is never without the book in his hand and who follows the regulations in the book to the letter. Concrete specifications are written as a guide and should not always be followed to the letter. It is up to the inspector to use good judgment in individual cases.

If a load of concrete is delivered with a slump of 4½ in., and the specifications call for a maximum slump of 4 in., the load should *not* be rejected. A sensible inspector would caution the ready-mix driver to be more careful on future deliveries or the load may be rejected. Of course, it the concrete has a slump far in excess of requirements, there should be no hesitation in rejecting it. Failure to reject such a load would be an open invitation to further infractions, along with being a dereliction of duty.

Variations in the air content of concrete are noticeable to a minor degree with slump change. Regardless of the type of structure being constructed, these minor variations should be overlooked as long as they are within reasonable limits. In hot weather, or if trucks have been mixing any length of time, the concrete temperature rises and can cause a sharp drop in air content. When the concrete being placed is exposed to the elements and subjected to freezing and thawing, acceptance of such a load is doubtful. Air contents below 3½ percent in concrete exposed to severe weather conditions will not provide the necessary protection. In order to reduce the possibility of lower-than-necessary air contents, keep the air on the high side of the specified range at the start of the job. Ordinarily, air percentages will be higher in the morning than later in the day. It is best to keep the concrete supplier informed if the air content is dropping so that he can increase the dosage of the air-entrainment agent to maintain the higher range limit.

Most contractors want acceptable concrete—that's what they are paying for. It is they who will be held responsible for faulty concrete, and they respect an inspector who is knowledgeable and conscientious. Such an inspector is of great value on a job and can save the contractor a lot of headaches by controlling the quality of the concrete placed.

There are few concrete masons or laborers who will not complain when placing low-slump concrete. The constant demand for "more water" can frustrate the inexperienced inspector. Laborers may throw down their shovels or rakes and walk off the concrete in apparent disgust. Generally this is an act designed to encourage the inspector to permit a higher slump. However, there are occasions when their complaints are justifiable. Where the permissible slump is 4 in., there is no justification for the inspector to insist on a 2- or 3-in. slump. The contractor, or his representative on the job, has a right to demand the allowable slump limit, and his request should be recognized.

There are occasions when a slump in excess of the maximum specification should be permitted, such as concrete placement where steel reinforcement does not present a good condition for the placement of a stiff mix. Where such conditions are expected, the contractor should request a mix design with the slump necessary and be willing to pay for the difference in cement factor cost to maintain an equal W/C ratio.

The inspector can be guided by jobsite cylinder test reports on just what excess slump he can accept if conditions warrant. An increase in slump of 1 in. will decrease the 28-day compressive strength by approximately 200 psi. If past strength reports indicate a comfortable margin of overstrength, the inspector may, when necessary, permit the placement of acceptable excessive slump. Conversely, if the past reports are just barely meeting required strength, he has little alternative but to reject the load.

INSPECTION OF LABORATORY

Regardless of who retains the testing laboratory, it is wise for the project engineer, the contractor, and the prospective concrete supplier to visit the laboratory chosen to inspect its facilities. Is the curing room adequate to accommodate the amount of cylinders received during the busy season? How do they cap their cylinders? What type of machine is used to test the cylinders in compression? What are the dates of the unstripped cylinders not yet placed in the curing room? Are they days old or longer? Were they made by the testing personnel or delivered for testing by someone else?

According to ASTM C 31, "Specification for Making and Curing Concrete Compressive and Flexural Strength Test Specimens in the Field," cylinders should be stripped and placed in a curing room at controlled temperatures the day after casting. This regulation is not always followed, however; for example, when the cylinders are made on a Friday, they are often not picked up until Monday. Generally, cylinders cast on a Friday will reflect lower strengths than those returned to the testing laboratory in 24 hours, stripped, identified, and carefully placed in the curing room.

INSPECTOR'S TESTING ABILITY

Every inspector should be familiar with the techniques of performing slump tests and air tests, and the procedure for molding cylinders. They should know, for example, not to sample concrete from the first 2 yards or the last yard of concrete from a ready-mix truck. All too frequently, we see jobsite cylinders being picked up and carelessly thrown into the back of a pickup truck. A blanket on the bottom of a pickup truck is not sufficient protection for newly cast cylinders. They can jounce around and bang into one another or into the sides of the truck. Cylinders treated in this manner are bound to reflect lower strengths than would normally be expected. It is preferable to transport cylinders in wooden cradles lined with foam rubber to prevent them from possible injury.

Inspectors should improve their kowledge by reading and attending technical lectures. There are, and will continue to be, books and articles on concrete subjects which are pertinent to all members of the construction industry. Publications by the Portland Cement Association, individual cement companies and admixture manufacturers, the American Concrete Institute, the National Ready Mixed Concrete Association, *Modern Concrete* magazine, and *Concrete Construction* magazine are readily available to improve a knowledge of concrete. Read them and file pertinent data for future reference.

Many people holding responsible positions in the concrete industry began their careers as technicians and/or inspectors. Contractors, engineers, superintendents, testing laboratory owners, managers, manufacturers, and concrete producers received their basic training in the testing and inspection field.

13 / THE CONCRETE PLANT

EARLY HISTORY OF READY-MIX

In the early stages of the history of ready-mix production, the concrete plant was usually a many-storied structure with the concrete truck stop at ground level, the batching floor above that, and the aggregate holding hopper forming the roof of the building. A narrow discharge chute below the weigh scale on the second floor carried the batched aggregates into the concrete truck below. The cement was added by bag, by hand, into a hatch at the top of the horizontal truck barrel. Generally, the bag cement was stored on a platform erected at the level of the top of the truck, so that the contents of bags could more easily be deposited into the concrete truck. These plants were known as *transit mix* plants.

The fine and coarse aggregates were stockpiled separately next to the plant and loaded into the storage hoppers by crane. In metropolitan areas, where plant sites were accessible by waterways, aggregates were brought in by barge and either loaded directly into the plant hoppers or, to reduce demurrage charges, stockpiled for eventual use.

Today, cement is commonly purchased in bulk form and delivered in special railroad cars or by truck. Cement storage bins or silos are constructed as part of, or as an adjunct to, the aggregate batching plant. Although the first concrete plants were erected as permanent structures, there are now many types of mobile plants which can be delivered to a site, erected in a few days, and then easily dismantled and moved when they are no longer needed.

MODERN-TYPE PLANTS

The low-profile mobile plant in Figure 13-1, a one-unit cement/aggregate system, is easy to erect. With a capacity to produce 110 yards per hour, it is a good starter or second transit mix plant.

A compact, mobile central mix-type plant (Figure 13-2), with a 4½-cu-yd batch, is also available for, say, remote areas.

Like today's cars, concrete plants are available in many shapes and sizes either to meet the standard production operation or to serve special purposes. On large projects, some distance from a permanent plant,

Figure 13-1 Low profile plant erected to supply dry batch mixes for highway construction. (Source: Pit and Quarry Publications.)

Figure 13-2 Mobile central mix plant. (Source: Pit and Quarry Publications.)

many concrete suppliers erect mobile plants that pay for themselves in reduced trucking costs.

There are basically two types of plants used in the ready-mix industry. One type, generally referred to as a *central mix* plant, deposits preweighed materials into a mixing barrel, adds water, and discharges the mixed concrete into the delivery truck below. The second type is the *transit mix* plant. At this plant the preweighed materials are fed into the delivery truck in their "dry" state and delivered to the customer's jobsite, where water is added and the concrete is mixed to produce the required slump.

There are strong supporters of both types. The central mix plant owner feels his method preshrinks the concrete, permits better control of slump and air contents, and allows additional yards to be hauled. Indeed, the last claim is valid. As ready-mix trucks have two different capacity ratings, one for the agitation of premixed concrete, and another, *lower* capacity rating for the transit mix truck, the permissible load carried by the central mix plant can be 2 to 3 cu yd additional. This extra payload is certainly an important factor to the concrete producer.

TRANSIT MIX PLANTS

The owner of a transit mix plant, although he may admit to some of the advantages of the central mix plant, tends to think more of its possible disadvantages. Being aware of the laws regarding weight restriction, strongly enforced in most areas, he fears that the additional weight of the extra yardage carried will draw overweight fines. Also, if the delivery truck has a long delay in traffic, or a flat tire, or a mechanical breakdown on the way to the job, the premixed concrete may stiffen. Especially in hot weather, he fears the rise in concrete temperatures will reduce the slump and air content of the load. In addition, he is concerned about the time spent, at the end of the day, in cleaning out the mixing barrel at the plant. If the plant mixer is not well maintained, it can break down, and the plant could be out of business.

CENTRAL MIX PLANTS

By contrast, the central mix plant operator considers the transit mix plant obsolete. He will agree that his trucks are fined for being overweight occasionally, but so are the other guy's trucks. He feels the central mix plant controls the slump entering the truck and doesn't leave it to careless or inexperienced truck drivers (who may soup up the concrete), eliminating the possibility of a rejected load. Further, a job may be serviced with one delivery rather than two. Truck wear and tear is reduced by not having to mix the concrete load during transit or on the job.

Naturally, the owners of each type of plant think their system is best. They are familiar with their operation and reluctant to change. However, there are times when problems arise that make the owners of either type of plant wish for the other kind. The obvious solution is to incorporate both systems in a plant. There are an increasing number of new plants being erected with a built-in flexibility that permits the use of both systems.

BATCHING

Regardless of the system used, the cement, sand, and coarse aggregate must be carefully weighed to conform to the mix design being batched. Aggregates are weighed in one scale, and the cement separately in another. Some batchers weigh these materials by opening and closing the gate at the bottom of the aggregate storage hopper by hand levers. A beam scale is set by hand, and when the scale indicator shows the desired weight is achieved, the material gate is closed by hand.

Today, many of these plants have converted to semiautomated weighing systems. Storage-hopper gates are opened and closed by buttons which control compressed-air cylinders, which in turn motivate pistons attached to the gates. The cement scale beam can be preset: the gate can be opened by button pressure, so that the gate will close automatically when the scale beam has risen to the correct weight. This improvement has permitted the batcher to concentrate on the weighing of the aggregates only and has thereby increased production.

The majority of today's plants are fully automated. A card programmed with the desired mix weights is inserted into a console, and all materials, including total gallons of water, are accurately batched. A push of a button activates a probe in the sand storage hopper to record moisture content. A dial on the console is turned to the proper moisture reading on the moisture compensator; at the same time the machine adjusts to the corresponding material weight automatically. All mix designs are made with SSD weights. The size of the weigh hopper will determine the number of cubic yards that can be batched automatically.

AUTOMATION

Early fully automated systems were large consoles with complex wired relays that usually demanded the service of the manufacturer's repairman to correct problems. Downtime was costly since there were few qualified repairmen and they were not always readily available. In many cases the manufacturer was some distance from the installation, and after the initial warranty ran out, the cost of the repairman's travel time and expenses often was prohibitively high.

The newer types of automated systems (Figures 13-3 and 13-4) are minute compared to early models. Solid-state systems permit accurate consoles no bigger than some television sets. They are reasonably trouble-free, and problems that do arise are corrected quickly and easily. The manufacturer, if not local, will commonly make an arrangement with a local repairman who can usually locate and correct the problem with little or no downtime, since all these systems have manual controls for emergencies.

Figure 13-3 Automated systems are still in use after many years service. (Source: Pit and Quarry Publications.)

Many other low-cost features are available in newer models. Standard features include the ability to handle 50 or more mix-design cards. Most systems today have a printout recorder which automatically reports the material weights batched, the water, the date, the time, and the truck number of each batch produced.

One interesting extra feature on these models is the memory recall. When used at the end of the day's operation, the recall device prints out the total pounds of cement and aggregates used during the day's production. This is a worthwhile feature for accurate inventory control. For the ready-mix producer the recorder has also proven very helpful in handling customer complaints. Any questions regarding batch quantities for any particular mix can easily be checked. Many state and federal agencies require a concrete plant to have such a recorder if it is to be used as a possible project supplier.

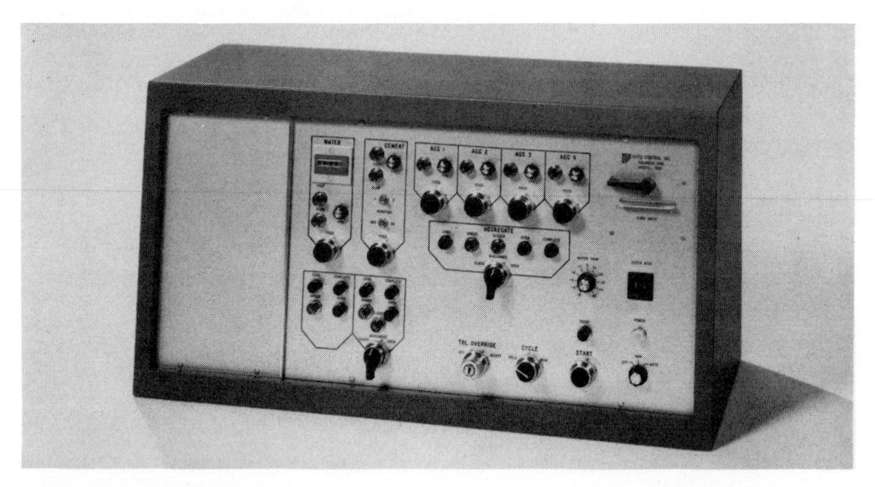

Figure 13-4 One of the newer automated-type systems that basically accomplish everything the older and bigger models do. (Source: Pit and Quarry Publications.)

MATERIAL DELIVERY AND STORAGE

Cement is delivered to the plant either by truck, from which it is blown through pipes to the storage silos, or by railroad car, from which it is carried, by screw conveyor, to the storage area.

Aggregates are delivered to the plant by truck, barge, or railroad car. They are dumped directly into plant holding bins from ramps, are dumped through grates on grade which deliver the material to bins by conveyor belt, or are stockpiled for later use. Railroad gondola cars are emptied by crawler crane or dumped through openings between the railroad ties into hoppers with conveyor belts beneath to carry the aggregates to stockpiles.

The method used for handling aggregate supplies is often determined by the plant space available. Certainly no plant site should be selected that does not permit space for large stockpiling of aggregates. Ground storage is protection for those extra-heavy production days and insurance in the event of an aggregate manufacturer's plant breakdown and/or winter shutdown.

Plant storage bins are fed by crane, which feeds material directly into the bins from the stockpile. Bucket loaders are used to feed a small hopper from which a conveyor transports the aggregates into the plant storage bins. There are systems in which stockpiled materials are divided by partitions with a conveyor below the stockpile, permitting the opening of a gate to select the aggregate required. There are concrete silos which have a conveyor system below for delivery of stored aggregates to plant bins. Each of these is an excellent method for storage in hot or cold weather.

Aggregate storage bins at a concrete plant are usually built in rectangular shapes and taper down to the size of the gate opening above the weigh hopper (Figure 13-5). A minimum of four separate bins is generally necessary, one for sand, one for ⅜-in. aggregate, one for the coarser aggregate, and a fourth for special aggregates, such as lightweight material. As there is more coarse aggregate used in a cubic yard of concrete than sand, this fourth bin, when it is not needed, can be used for additional coarse aggregate.

Figure 13-5 Concrete plant with aggregate storage bins above conveyor that supply plant. (Source: Pit and Quarry Publications.)

The bins should be emptied periodically to check for partition wear. Many a bony-load complaint can be traced to a hole in the divider partition which permits the entry of coarse aggregate into the sand bin when the level of the sand is below the level of the coarse aggregate. Steel rusts when constantly damp, and the abrasion of the aggregate will soon wear through even a diamond-plated material divider. Concrete plant owners have discovered that lining the bins with old conveyor belting reduces this normal wear. When a hole is discovered in the partition, bolting a patch of belting over the hole will provide a quick and serviceable emergency repair.

WEIGH SCALES

The weigh-hopper suspension scale is directly below the aggregate holding bins. The bottom bin gates for loading the aggregates into the weigh hopper are usually in a line, making it necessary for the weigh scale to be rectangular. This is not always the case: some aggregate gates are dual, permitting the weigh hopper to be square, or, as some prefer, constructed in a hexagonal or octagonal shape. All weigh hoppers funnel to a small opening at the bottom to discharge the aggregates into the concrete truck below.

The damp fine aggregate has a tendency to hang up in the angles of the square and rectangular scale; it is commonly dislodged by the use of a vibrator. Some plant owners, if the weighing operation permits, prefer a six- or eight-sided weigh hopper with a rounded cone tapered to the discharge gate. These vastly reduce aggregate hang-ups and eliminate the delay which vibration would create. Those with the square- or rectangular-shaped hoppers can reduce hang-up by welding straight pieces of steel over the corner angles.

Beam scales, one beam for the fine and one for the coarse aggregate, reflect the weights of material in the weigh hopper. On the other hand, the dial face scale eliminates constant adjustment, reduces the possibility of error, and permits faster loading. The total coarse aggregate weight for the batch is usually weighed first, and the additional weight of required sand is added to this first weight until the combined total of the batch is reached. The coarse aggregate is weighed first to allow quicker discharge of the batch into the delivery truck. If the loading procedure is reversed, the damp sand tends to pack, slowing production.

THE QUALITY CONTROL ROOM

Adjacent to the batching operation there is usually a small room for the testing of aggregates by plant personnel, and/or for the use of plant inspectors. The normal small laboratory is equipped with a set of fine- and coarse-aggregate sieve sizes for gradation tests, a hot plate for drying sand samples, and a small 1000-g scale. A larger platform scale with a 100-lb capacity is ideal for weighing coarse-aggregate samples.

One might include, if only for personnel use, a supply of a 3% solution of sodium hydroxide and medicine bottles to use in testing for deleterious material.

LEFTOVER CONCRETE

The disposal of leftover concrete presents a serious problem to the concrete producer. Because of more stringent conservation and ecology laws, he can no longer dump the unused concrete wherever he chooses, and a surprising number of cubic yards of concrete are returned to a plant in the course of a year's production. The plant owner must now provide ample space, away from streams or other water sources, for the daily deposits of unused concrete. Most producers, unless they have unlimited waste space, build, or purchase, a concrete washout system.

Leftover concrete is dumped into a concrete separator filled with water which automatically separates the cement and aggregates and stockpiles them individually. The cement, when dried, is no longer active but can be used for a multitude of purposes. The clean sand and coarse aggregate can, of course, be reused in concrete production. All water is clarified internally and can be used for rinsing the barrel or for fresh concrete.

The size of the washout system will vary according to the number of trucks in the fleet. Most of them will permit the simultaneous dumping of 3 or 4 trucks and will have the overall capacity to handle 30 or 40 partial truckloads. Although few concrete producers have accurate figures on the cost of purchase and operation of these units, all agree that the reclaiming of the aggregates will eventually cover whatever expense is incurred.

14 / THE CONCRETE PRODUCER

The concrete plant is the producer's factory, which in itself is only one essential part of his production. He must have raw materials delivered to his plant in order to manufacture his product. He must have a fleet of ready-mix trucks to deliver the finished product to his customers. Further, he must know how to produce quality concrete if he intends to remain in business.

TRUCKS

Cement, which is the most expensive ingredient in the manufacture of concrete, is delivered in special cement trucks, owned or leased by the cement company. The fine and coarse aggregate is generally delivered in 10-wheel or tractor-trailer trucks. Usually, these trucks are owned by the concrete producer.

The high cost of these trucks and of their maintenance (the cost of employing one or several mechanics and of maintaining an inventory of spare parts) plus the high cost of fuel for their operation all form a major part of their total operation expense.

The selection of a ready-mix truck is an important decision for the producer. It is usually preferable to keep a fleet of the same truck make to eliminate the need for keeping a variety of spare parts for different makes. Mechanics become familiar with one model and perform repairs in shorter time. In addition, the ready-mix drivers will feel more at home in the cab when and if they have to change trucks.

Originally all concrete truck mixer barrels were horizontal. The concrete materials were introduced into the mixer by gravity feed through an opening on top of the barrel. The loading of the material had to be halted occasionally, and the barrel then rocked back and forth, before the full capacity of the load could be introduced. Shown in Figure 14-1 is a 1925 3-yd-capacity model of this type.

The horizontal type of mixer barrel did a good job of mixing concrete and, in time, was built to carry as many as 15 cu yd of transit mix concrete. Many of these mixers are still around today, but the more

Figure 14-1 A 1925 model concrete truck with horizontal mixer barrel. (Source: Pit and Quarry Publications.)

popular "high-boy" type of truck, shown in Figure 14-2, is the general standard.

Materials are loaded into the hopper at the rear of the truck and drawn into the barrel in the loading sequence. The mixed concrete is discharged by reversing the rotation of the drum.

The horizontal barrel required the driver to remove the loading hatch for receiving materials, and to close it when the batching operation was completed. A door at the rear of the truck had to be closed for mixing the concrete and opened for discharging it. Buildup at the door was common, and constant maintenance of this section was necessary to prevent such buildup and to allow firm closing.

Loading time is less with the high-boy than with earlier trucks. Because the discharge end of the truck is higher than the mixing section of the barrel, no rear door is required. The additional height at the discharge point permits easy delivery into foundation forms.

Some high-boy trucks require that the driver operate the mixing, the addition of water, and the discharging of the concrete from controls at the side rear of the truck. In other models the driver remains in the truck cab and conducts all operations by controls on the dashboard. An optional feature is a slump indicator, installed in the cab, which tells the driver the slump of the concrete in the mixer barrel.

Figure 14-2 Concrete truck of the high-boy, rear-end discharge type. (Source: Pit and Quarry Publications.)

Figure 14-3 Concrete truck of the front-end discharge mixer type. (Source: Pit and Quarry Publications.)

Gaining in popularity is the front-end discharge mixer shown in Figure 14-3. This type of truck eliminates the necessity of backing the loaded truck into position to discharge its load. The driver faces the discharge end of the chute and can control the flow of the concrete being unloaded.

When unloading is finished, *backing out empty* is easier than *backing in loaded.* The one minor problem of constant cleaning of the truck cab, due to concrete spillage from the discharge chute, would appear to be a small factor when the advantages are considered. Most concrete customers prefer this newer type of truck.

PRODUCER'S RESPONSIBILITY

In all too many cases, rightly or wrongly, any concrete problem that develops subsequent to delivery is blamed on the concrete producer. Of course, the producer could point out that ASTM C 94, "Specifications for Ready-Mixed Concrete," relieves the supplier of responsibility for his concrete after it is discharged from the truck. The producer could simply deny responsibility for problems which may have resulted from the poor workmanship of the contractor. However, if he followed this practice, he would soon be without customers. Moreover, the complaint, regardless of how it is put, is frequently a cry for help.

PROBLEM SOLVING

Problem solving is therefore a necessary service of the concrete producer to his customer. Along with his knowledge of plant operation and his ability to design and control the different mixes, the producer must himself be, or must have on hand, a technician with a knowledge of all things concrete.

This knowledge cannot come from personal experience alone, valuable as that is, but must be gained by additional help from other professional sources. Attending concrete seminars or concrete instruction courses held by associations and concrete product manufacturers are most helpful. Publications of specific problems, such as those issued by the Portland Cement Association, the American Concrete Institute, the National Ready Mixed Concrete Association, or individual cement companies and admixture producers, are readily available free or for a small fee.

Copies of these informative publications should be kept in a file cabinet for ready reference. Even if the data in some of these articles or pamphlets do not seem immediately useful or fully understandable, the publication should be filed for future reference. As given problems arise, they may make a lot of sense then.

Many concrete technicians and salesmen carry copies of such publications in their cars and show them to customers who have a problem related to an article. There is a tendency occasionally for the customer to distrust a salesman's explanation of the cause of the problem on the assumption that he is merely defending his employer. When shown a publication by an authority in concrete, the customer is more readily and thoroughly convinced.

Frequently, the customer knows what caused the problem and has really complained in the hope that the producer's representative will offer a cure for the fault. When the person answering the complaint has the solution to the problem, that customer is not ever likely to switch his allegiance to another company.

A PRODUCER'S CONCRETE SEMINAR

In the off-season it may well be helpful for the producer to have a 1- or 2-day concrete seminar, conducted at his plant or at a nearby motel. If there are no personnel of the ready-mix company competent to conduct the course, a testing laboratory can be employed for the purpose. Admixture manufacturers have a wealth of talent on their staff capable of conducting a seminar and will often do it just to advertise their products.

Invite architects and engineers as well as contractors, although not necessarily at the same time, to these free sessions. Producers who have conducted this type of educational program have increased their own knowledge, improved public relations, and accomplished an increase in sales as well. Just as important, through questions asked by the participating audience, a better understanding has developed of what is needed or wanted by the industry.

Before running these sessions, the producer and his technical assistants should ascertain what their customers know or don't know, what they want and don't want, and what they like or don't like. The scheduling and structure of such sessions might follow this pattern: Copies of interesting and useful literature should be handed out at the beginning of the session (at about 8:30 or 9:00 A.M.). Free coffee and a light snack should be provided before the session starts, with a buffet lunch provided later which might include one drink. Finally, an open bar may be included at the end of the seminar at about 3:00 or 3:30 P.M.

The overall intent of such an educational and social event is to show that today's concrete producer is not just a trucker of concrete.

HANDLING CONCRETE PROBLEM CALLS

It is possible, although highly improbable, that in the future a panacea will be introduced to cure all concrete problems. Until that time the construction industry can only strive to eliminate or reduce the known causes that create unnecessary problems. Not all concrete customers are competent concrete contractors. Others, while reasonably knowledgeable, are careless. Still others, though they may well have the best of intentions, lack the experience and ability to handle the delivered product.

Regardless of the ability of the contractor, when a problem does arise, it is generally the concrete supplier that is called upon for an explanation. He, or his technical representative, may never have visited the actual jobsite when the concrete was placed.

However, the concrete producer often overlooks the one person in his operation who is on every jobsite every day—his ready-mix driver. While these drivers are usually neither college graduates nor concrete technicians, they are intelligent and observant about what goes on at a jobsite. They know good placing practices from bad and can tell the competent customers from the incompetent. They are an excellent source of information when a concrete complaint is registered. More on the training of drivers can be found in Chapter 15.

Every driver is in a sense a "salesman" since his personality and his behavior on the job reflect well or ill on the company he is employed by, and thereby influence sales. However, the initial contact with a customer is made by an actual salesman, on whom the concrete producer relies heavily for a successful operation.

Due to the nature of the business, the salesman must wear many hats. He must be a public relations man, spending a good part of his day visiting customers in the field. He must be the company representative who handles complaints and who must therefore have a good technical knowledge of concrete. He must be a politician who can handle any problem with tact and understanding. He must be mentally mature and able to hold his temper in the face of a customer's anger and, often, exaggerated accusations.

THE DISPATCHER'S ROLE

A good salesman is a valuable asset, but the dispatcher of a ready-mix operation is unquestionably the key figure in the concrete producer's business. On him, more than on any other single member of the concrete plant's personnel, the reputation and success of the concrete producer depends.

His job is extremely demanding: One can only wonder, having watched a dispatcher work for a whole busy day, why he returns the next morning. He usually arrives at the plant before anyone else, to receive early customer calls. Usually these early callers are customers who forgot to call in their orders the previous day. They become irate when no delivery trucks are available, and they blame the dispatcher. A good dispatcher, who knows his customers, will usually placate the caller and promise to squeeze him in during the day. Usually he does, in spite of a full schedule.

The dispatcher receives the butt of all complaints. Whether they are justified or not, he must remain pleasant and understanding. He must take a concrete order and give the impression that the order, regardless of the quantity, or the manner in which it was given, is appreciated. He must write delivery tickets, juggle his trucks for prompt delivery, answer phones, and give directions to drivers (and keep one ear on the intercom if the delivery trucks are radio-equipped).

He must constantly communicate with the batcher to inform him of which truck is to be loaded and with what number of yards, what mix, and what admixtures, if any. When the concrete order is COD, he must calculate the charge, computing the proper tax and invoicing it on the delivery truck. Any mistakes on his part in any of these operations can bring down the wrath of the customer or the plant owner.

Along with the routine duties of his job, the dispatcher must contend with unexpected truck breakdowns, the unforeseen extra yard or two needed to complete a pour, an unanticipated delay of an hour or more on the jobsite, the cancellation of an order, or the request for delivery later than originally specified. Any of these unplanned-for interruptions throws the carefully scheduled truck deliveries out of sequence and necessitates rearrangements which require skill to manage successfully.

He is often the last man to leave the plant since he must stand by either to receive late orders or to make sure a late delivery will not require a repeat. A good dispatcher is a special kind of individual. He is dedicated and loyal and frequently poorly paid in relation to the time and effort he expends. He can be, and often is, the most important member of the management team.

THE CONCRETE PRODUCER'S ROLE

Certainly every business involves stress and worry for its owner; but the concrete producer seems to have more than his fair share. There are either too many trucks or too few relative to the volume of orders. There is the threat of bankruptcy with a downturn in the economy. There is the unpleasant routine of getting up at 5:00 or 5:30 A.M. to be at the plant before the daily operation beings to ensure that deliveries scheduled for 8:00 A.M. are loaded by 6:30 or 7:00 A.M. depending on the length of the haul. There is the concern of the inevitable breakdown of a ready-mix truck late in the day on a jobsite that requires the dispatching of some piece of towing equipment. Frequently, the truck gets mired on the job because the customer has not anticipated the need for a stable subgrade to handle the truck's load.

Pulling a mired truck free is often a difficult and expensive operation for the producer. Rear ends are damaged, clutches burned out, axles and bumpers broken. Often the contractor has washed his hands of the problem and left the site, leaving the supplier with no place to deliver the load. This means lots of trouble and expense, and no sale.

The production of concrete is a difficult and expensive operation; it takes a special kind of person to enter and then stay in the business. He must be a man who likes challenge and can stomach aggravation. Yet, although he may bewail the fate that ever got him involved in such a problem-filled industry, the few who do persist in the ready-mix business never leave when given the opportunity. Perhaps they just like concrete.

15/HELPFUL HINTS FOR THE CONCRETE PRODUCER

ESTIMATING DAILY PRODUCTION

Every concrete producer has, or should have, some method of determining the relative value of his daily production. The total number of yards per day, or yards delivered per truck per day, is not an adequate index. The best criterion for daily production is the yards per driver-hour. This method takes into consideration the long hauls, short loads, waiting time, overtime, breakdowns, and delays in loading.

A few years back a survey reported 2.7 yards per driver-hour as the production average. This average took into consideration those plants located in or near metropolitan areas and those some distance from a city or town. By keeping a daily record of yards per driver-hour and by subsequently reviewing his work sheets, the producer can determine the reason or reasons for falling below *his* average. Were too many trucks ordered in for the day's delivery? Was there too much time lost in washing out or loading? Were certain customers taking too much time to unload the truck? Was the dispatching operation poorly planned? These and similar questions can be answered if such records are kept and studied.

UNLOADING TIME

Touring the country, one finds that producers in different areas apply different standards in computing charges over and above the initial cost of the concrete. Some limit unloading time, calculated from the starting time of the unloading to its finish. Some limit this time to ½ hour for loads of 7 yd or less, others to an hour for the same quantities. On larger deliveries some producers allow 1½ hours to unload. A few do not charge at all for this "waiting time" period.

The "no charge" practice is difficult to justify since the cost of the driver and the truck is often 25 percent or more of the net selling price. Agreed, the customer who usually discharges his delivery in record time should not be charged for an occasional delay, but the fact that the producer's competitor may not charge extra should not sway him to exempt all customers from charges for waiting time. True, there may be a loss of some customers when the practice of "waiting-time charges" is instituted, but volume does not necessarily mean profit.

OVERTIME CHARGES

Overtime, that period of the day when the driver works beyond his standard 8 hours, should be charged to the customer, unless it is part of a large pour that began early in the morning. Moreover, overtime should be charged not only for the driver's time—this alone would not compensate for the cost of the truck—but also for the batcher standing by in case the customer runs short, and for the mechanic who stands by in case of a truck breakdown. The overtime charge will drastically reduce those late-afternoon delivery orders.

SHORT-LOAD CHARGES

An additional charge for "short loads" has become standard practice. There was rarely an extra charge for short-load deliveries when the older and smaller-capacity ready-mix trucks were used. But today's larger trucks, capable of carrying 10 cu yd or more, are expensive to purchase and more costly to operate. To send trucks out with deliveries of low yardage, unless used to complete a larger order, is a losing proposition. A producer must determine his minimum-profit load and charge extra for yardage delivered under that load.

Some producers, for example, have a flat additional charge for loads under 5 yd. Under scrutiny, in lieu of today's high operating costs, this system is a poor arrangement. More and more producers have gone to an escalating cost scale depending on the number of yards delivered in the short load: so much extra for 1 yd, less for 2, and so on. Regardless of the extra charge, it is not profitable for the producer, during the average day's operation, to accept the occasional short-load customer until the busy part of the day's production is over.

The average dispatcher likes to have an extra truck or two each day to service the customer who calls in an order on the morning of the day he requests concrete delivery. Some producers make a practice of ordering one extra truck for each five or six trucks needed to accomplish the known production. Frequently these extra trucks fulfill a need and deliver a reasonably good day's production. Just as often, however, they deliver one or two loads and do nothing more than lower the daily production average of yards per driver-hour.

LONG-DISTANCE CHARGES

The concrete producer must charge extra for distance hauling. Obviously he cannot charge the same price for concrete delivered 15 mi from the plant as for a delivery 5 mi away. Many concrete suppliers use a zone system. Using the plant as the center point, they draw circles to locate delivery sites within a definite radius, thereby determining zone charges. This method is no doubt easy to implement but not completely true to circumstances. In fact a shorter-distance delivery in one direction may, because of traffic delays or the like, be more time-consuming and costly than a delivery twice the distance in another direction. This raises the question of whether *distance* or *time* should be the prime consideration in determining charge.

So many factors come into the picture that the answer to distance or time must be made by the individual producer. Starting with the known fact that delivery cost is approximately one-third of the net selling price, the producer must consider the location of the plant (whether in a rural or metropolitan area), the hourly wage of the driver, and the cost of fuel in determining the best choice. Most producers have agreed that time is the most important element in profitability. In any case each job should be priced on the importance of the customer, the total yardage involved, the availability of trucks, the current business outlook, and the time of the year.

THE WINTER MONTHS

In cold weather areas concrete production during the winter months is expensive and usually winds up an unprofitable operation. Many ready-mix plants operate at a loss during the cold weather months. The water and aggregates must be heated and the plant itself maintained at a minimum of 50°F to keep pipes and scale pots from freezing. Further, the aggregates themselves must be kept from freezing both in the holding bins and in the stockpiles. The usual added charge for winter concrete of 1 or 2 dollars per yard is never enough to compensate for the high cost of the heating fuel needed for production. In addition, although snow, ice storms, and extreme cold weather may halt production for days or (often) for a week or more, the concrete supplier must keep his plant and materials heated 7 days a week.

When there is a large project underway, the concrete producer must stay open during cold weather to service the job, unless otherwise stipulated in the contract. In areas where cold weather temperatures are extreme, the small producer closes shop. In more temperate zones the concrete plant operator feels he must remain open to serve the occasional customers, or lose them to his competitor.

Recently, however, concrete plants that have previously remained open in the past have closed their doors from the middle of December through March, and have profited from the shutdown. This provides an opportunity for conducting needed repairs and maintenance on the plant and trucks while not suffering much, if any, customer loss.

In cold weather, in areas where there is more than one supplier, the suppliers have gotten together and have agreed to shut down their plants on alternate years, letting one plant service the area. Although operation is not highly profitable during cold weather, at least service is provided to the customers of both companies, and the added business to the plant that remains active helps to defray the cost of winter operation.

THE READY-MIX DRIVER

Any plant downtime provides an opportunity for a brief training course for drivers. Before spring startup, or on a rainy day in the early season, one of the plant personnel familiar with the proper methods of field testing could conduct the training session. Some producers employ the services of a local testing laboratory to conduct the course, after hours, with free beer as an added attraction.

The slump test should be the first item for instruction at such a session and should be taught not just abstractly but by actually using concrete. Every driver should know the proper procedure for testing slump as specified by ASTM C 143, "Test for Slump of Portland Cement Concrete." Too many slump tests taken in the field are conducted haphazardly, and often a concrete load is rejected on the basis of the result of such an improperly made slump test.

After the method of taking slump tests is completely understood, the session should cover the proper procedure for the molding of concrete cylinders and the testing of air. Improper procedures in the molding of cylinders, and in their care after molding, create more low-strength problems than any other single factor in the testing procedure. The cylinder mold, usually made of cardboard and coated with wax, is 6 in. × 12 in. The bottom is usually made of metal. The cylinder, as in the slump test, should be placed on a firm level base (wood is fine) and the concrete placed in three equal layers, each rodded 25 times with a ⅝-in. bullet-nosed rod. The use of anything other than a bullet-nosed rod is unacceptable. When the second and third layers are rodded, the rod should penetrate into the concrete below.

The concrete sample for both the slump and/or cylinder tests should never be taken from the first or last two yards of the truck load. Any slight nearby vibration may affect the molded cylinders and will surely affect the result of the slump test. The vibration of a concrete truck during mixing or unloading is enough to increase the slump of the concrete sample. Collect the concrete in a premoistened wheelbarrow and take it to a quiet spot for the test. Be sure the base of the slump cone is firm. A wobbly base can strongly influence the result of a slump test, and it is all too often overlooked in a field test and should be watched for.

When testing for air contents of the concrete, an air-pressure meter is the most accurate method to use. If the equipment is well-maintained and the operator is familiar with its performance, the result will be accurate. Many inspectors use the small Chace air meter, which is reasonably accurate in the normal slump range of from 3 to 6 in. When using the Chace, only the concrete mortar is sampled, isopropyl alcohol is added, and the glass vial is gently agitated until all the mortar is in solution. The alcohol dissipates the air in the mortar, and when the finger is removed from the vial opening, the markings on the long stem indicate the air content. However, unless the alcohol is at least a 91% solution, the test is unreliable.

A good knowledge of the three test procedures above is most helpful to the concrete producer if his drivers have had training in these areas. Drivers should be requested to note any improper procedures that they notice on a jobsite and report these facts to the plant office. One producer on the East Coast gives a bonus award to any driver whose report later helps to solve a concrete complaint. He contends that this incentive has proven most valuable.

In addition to such immediate economic benefits, the training course serves to increase the interest of the driver in a wider scope of the concrete business and creates a definite improvement in overall morale.

INDUSTRY PERIODICALS

Most concrete producers subscribe to industry periodicals. They should not only read these regularly but take note of articles which are likely to be of interest and use to their customers, make copies of these articles, and mail them out with the next billing. Even if many of these copies end up in the "circular file" unread, and even if the producer seldom gets an explicit "thank you," nevertheless the service is likely to be appreciated. Some customers will find the articles helpful in their operation; the articles will keep them informed of new concrete products or methods. In any event, they are more likely to remember the name of your company the next time they order concrete.

Naturally, not all concrete problems are created in the field. The concrete producer has his share of occasional problems that are sometimes difficult to trace. The dispatcher, the batcher, and the truck driver are all human and can make mistakes. Nor is the material manufacturer who supplies the producer free of guilt.

THE PROBLEM OF CHANGEABLE WATER DEMAND

One common problem in concrete production that the ready-mix driver will usually be the one to complain about is the problem of changeable water demand. It is possible that one delivery load will differ so much from another load by the same driver with the same mix to the same project that there will be a wide variance in water demand for the required slump, creating the potential of a *wet* load. Producers should be alert to the possible cause of such differences in delivery loads. A list of possible causes follows:

1. A faulty plant water meter
2. A worn hole in the partition separating the fine and coarse aggregate
3. A faulty admixture dispenser
4. An overloaded sand bin whose excess overflows into the coarse aggregate, or vice versa
5. A fine aggregate that varies in gradation and/or moisture content
6. An erratic aggregate scale

Of course, if the problem of changeable water demand exists with only one driver, it is usually the fault of the driver or his truck. If the complaint is registered by many, the cause of the complaint should be looked into promptly and the problem resolved.

A list of other possible complaints follows.

HIGH AIR CONTENTS

Normally, when excessive air contents occur, the driver is usually aware of the problem. The concrete looks "creamier," and a noticeably lower water demand for a given slump will be evident. However, the problem of high air content is not always evident until after the concrete has been placed. Concrete finishers may then complain of difficulties in the finishing operation. Surface blisters may appear after the first troweling operation. The concrete may be tacky and stick to the knee boards of the mason.

There is always the possibility, of course, that the concrete was finished with a steel trowel, rather than with a magnesium one. The steel trowel can seal the surface, preventing air from escaping through it. This entrapped air would then create a blister. When the complaint of excessive air comes as an isolated incident from a finisher, the finisher's operation might itself be questioned.

However, regardless of the source of the complaint, the producer should check the air contents of his concrete. If the air *is* much higher than usual, the producer might consider the following:

1. Is the air-entraining admixture dispenser accurate?
2. Has the gradation of the fine aggregate changed drastically?
3. Has there been a new shipment of air-entraining agent delivered?
4. Could a delivery of air-entrained *cement* have been received by mistake?

There is a simple test for checking air in cement. Fill a clean small-diameter cylindrical jar, such as olives come in, one-third full with cement. Add enough water to bring the sample to two-thirds full. Cap and shake vigorously. Then place the jar on a level spot and remove the cap. The sample will be topped with bubbles. If there is *no* air-entraining agent in the cement, the large bubbles will quickly disappear. If they remain for a considerable length of time, sometimes for hours, you can be reasonably sure the cement was air-entrained.

Occasionally, high air contents may occur without air-entraining cements or any other air-entraining agents. If a long drought has severely lowered the water table of the open water supply that is used to wash the fine aggregate, algae (that group of light-green one-celled plants that grow in warm ponds and small lakes) may be the cause of the phenomenon. Coating the fine aggregate, the dead algae decompose and give off gas that shows up in our test meter as air. Although rare, this phenomenon is not unknown; its possibility as a source of excessive air should be checked when there is no other obvious cause. Hose down a large sample of the sand in question and look for a light-green area of water after the sample has been washed, or check the condition of the source of the aggregate wash water at the sand producer's plant.

If the presence of algae is evident, the sand cannot be used. Put that sand aside and turn it over occasionally to let the entrapped gases escape. The sand will eventually be usable.

CONCRETE BALLING

Concrete balling will sometimes occur in the mixer barrel. These balls of various diameters contain cement and aggregate. They are not to be confused with lumps of hardened cement that usually are formed in the manufacturer's silo, the cement car or truck, or the concrete producer's holding silo. Cement will harden at any spot where there is an opening that will permit moisture to enter the cement storage area. Sooner or later, under vibration, that hardened cement will be loaded into the ready-mix truck.

Balling can be caused by the introduction of the cement into the delivery truck at the same time as the aggregates and water, particularly when the water is being heated. Balling seems also to occur more often in hot weather when the aggregates are warm and the cement hot. However, in many instances, the exact reason for the balling cannot be determined. If the balling condition occurs only in one or two trucks, one can suspect that there is a buildup in the mixer barrel of the truck preventing a good mixing action. There are times when balling develops suddenly, and without any change in the loading or mixing procedure, the condition disappears.

When balling occurs, for whatever reason, one can mix the concrete load at a very low slump to break up the balls. After several minutes of mixing at the low slump, additional water may be added to obtain the required slump. Mixing the load with a medium or high slump will not eliminate or reduce the size of the balls. However, when balling becomes a serious problem, a change in the method of loading the delivery truck should be considered. Loading the aggregates and a portion of the mixing water, stopping the mixer, and adding the cement separately, by gravity feed, will help considerably in preventing this condition.

RETARDED SET

Retarded set of concrete is a common complaint usually registered in the fall and spring of the year. We refer here not to the deliberate retardation brought about by the use of set-retarding admixtures, but rather to undesirable retarded set.

Every concrete producer has an additional charge for what is referred to as *winter concrete.* If the ambient temperature so dictates, the water and, when necessary, aggregates are heated to produce concrete temperatures above 50°F. In most cases there is a starting date and final date in between which this charge for winter concrete is enforced.

A bright, sunny spring or fall day does not mean the subgrade upon which the concrete is to be placed is free of frost. This is particularly so when the area for the placement of concrete is under cover. Concrete with a temperature of 70°F can drop to below 50°F in a half hour or less when deposited on such a base. With lower-temperature concrete, the drop in temperature is more drastic. In any case, concrete temperatures below 50°F will cause retarded set.

Agreed, it is the customer's duty to protect his concrete placement area from possible frost. This is an added expense to him, and often he would rather gamble that the ambient night temperature will be mild. However, he will usually complain to the concrete supplier if he runs into overtime to finish the retarded-set concrete.

The producer, in order to avoid such problems, must decide whether or not to supply hot water without charge. Generally the producer has only one water line at his plant to supply concrete-mixing water. If he supplies hot water to one, he must give it to all, not only to the floor customer but to all other customers whose projects may not necessarily need this feature.

Of course, if the producer's competitor heats his water at such times, without charge, he must supply the same courtesy. If his competitor does not, that should be an added incentive for him to do so. The additional business he will pick up will more than defray the added cost of heating the water.

SOFT SPOTS IN CONCRETE

Soft spots on newly placed concrete floors, although not a general complaint, do occur, and it is sometimes difficult to determine the reason for the phenomenon. These soft areas in an otherwise hardened surface are often the result of the addition of a set-retarding admixture in a mix placed with a low slump. Without sufficient mix water to dilute and disperse it equally throughout the load, the admixture can be concentrated in certain portions of the load and drastically delay the set at the spot they are unloaded. It may well be days, or more, before these soft areas will harden. Even if the slump is adequate for dispersal of the agent, soft spots can occur if the concrete has not been thoroughly mixed. Finally, coffee, soda, or any liquid containing sugar can create the same effect if deposited on fresh concrete, as can acids, chemicals, and similar agents.

On the other hand, the set-retarding features of sugar can occasionally be put to good use. For example, sugar can come in handy when a ready-mix truck gets stalled and cannot be moved for many hours. As long as the mixing drum can be rotated, the set of the concrete can be delayed indefinitely with the introduction of 5 lb of granulated sugar. Although the concrete itself will then be unsuitable for use, the possible expense of replacing the mixing barrel will be avoided, or at least the more probable expense for labor to chip out the set concrete from the barrel will be eliminated.

A high dosage of a set-retarding agent will achieve the same result as the sugar. When neither of these is readily available, flood the mixing drum with water.

SPLIT LOADS

The financial feasibility of split loads (two loads of concrete delivered to two different customers in the same truck) is questionable. When there are two small orders for concrete in the same general area, the decision of whether to send two separate trucks or send one truck to accomplish both deliveries must be made. The economic advantage on the one hand must be weighed against the problems that could develop with this less costly procedure.

Some customers, for example, are poor judges of the exact yardage they will need, particularly for several small placements. Their measurements are frequently haphazard, they may not take into account an uneven grade, or they may not measure at all but simply guess the amount of concrete needed. It is not unusual then for the first customer of a split-load delivery to use more concrete than he had ordered, leaving the second delivery short and requiring a second trip to complete the order. Not only is this second trip costly, but if there has been a delay on the first delivery, the second customer may receive a lower-quality concrete.

There is also the danger that the first customer wants high-slump concrete and the second wants a stiffer mix. Some suppliers add an extra ½ or 1 cu yd of concrete to the split load to assure that both customers will have enough. It may solve the problem of the customer using a little more concrete than ordered, but it does not solve the problem of the first customer wanting a higher slump than the second.

Many concrete producers have pretty much stopped the split-load practice. The system of charging extra for small yardage has eliminated the original reason for doing so. But the problem of the customer who underorders still exists. He may order 8 yd plus, and after that load is delivered, he will call up the concrete plant and order an additional 2 yd to complete the job. This necessitates a second trip with 2 yd of concrete, with a truck that had the capacity to deliver the whole 10 yd on the first delivery. Should this customer be charged extra for the second delivery of 2 yd? It is hard to answer.

This type of ordering occurs frequently when the concrete producer has upgraded his fleet to larger-capacity trucks than he previously had.

If his older trucks had the capacity to carry only 8 yd maximum, the customer who orders 8 yd plus cannot be fully blamed. The dispatcher, when taking the order, should request the total yardage needed, reminding the customer that the company now has larger delivery trucks.

If all concrete producers went over their accounts periodically, they might be surprised to find that quite a few of their customers are, in fact, not profitable. It is both prudent and fair to inform the errant customer, in writing, of his past behavior and of the financial burden it places on the producer.

WAITING TIME

Waiting time is, or should be, the time allowed a customer to unload a delivered yardage to avoid extra charges. Some producers allow a half hour and some an hour from the time the concrete is mixed on the jobsite to the time the truck is unloaded. Until recently, this was pretty much standard practice, but when a supplier uses the number of yards per driver-hour as a criterion, he will find this practice is not conducive to good production. On a delivery of 4 yd , an hour is far too much free time allowed, while, on the other hand, an hour may not be sufficient time to unload a 12-yd delivery. One way to correct the situation would be to allow, say, 8 minutes for unloading each yard of delivered concrete.

It is agreed that a producer, along with the sweet, must accept some of the bitter from his bigger, loyal customers, or on large construction projects. One cannot enjoy delivering several full loads of concrete to one customer, who may unload each truck in 15 minutes, and then expect to charge waiting time for one or two loads that took longer to unload than the usual allowable free time. Still, it is a good idea with any customer who takes an occasional short load, or has excessive waiting time, to note the infraction on the billing, and put the words "no charge" beside the amount of the penalty incurred.

READY-MIX TRUCK RENEWAL

Any additional charge for short loads, waiting time, overtime, and long-distance deliveries rarely fully compensates the producer for a truck's operation. This brings us to the question of how long a producer can keep a truck in use. The answer to that question involves weighing the cost of a new truck—which has soared out of sight in recent years—against the similarly escalating cost of repairs and maintenance. Likewise, the answer hinges on whether one measures the useful age of a truck in years or in mileage.

One large concrete and aggregate producer, rated among the top 10 in production and sales, employed a staff just to keep track of expenses for each truck in his fleet. The statistics he collected suggested that *6 years* was the maximum age for the profitability of a truck in normal operation. As a result the producer traded every ready-mix truck and all material trucks every 6 years. This well may be the intelligent thing to do with a large operation. Ordering trucks in quantity gives the producer the advantage of being able to dicker prices.

But what about the small producer who may only purchase one or two trucks at a time, every few years? He is in no position to bargain for price. The purchase of one or two trucks, even if badly needed, is often beyond his financial capabilities, and so he must make do with what he has, paying for extensive maintenance and repair parts to keep his trucks serviceable.

Many dispatchers and plant managers feel that the more trucks and drivers they have available, the better the service they can give. But it is impossible to serve all the needs of all the customers and make a profit. An increase in the number of trucks, even if it did somewhat increase customer volume, would not necessarily solve the producer's problem. The better solution requires that the whole operation be more efficiently run.

DISPATCHING

The dispatcher, for example, must know his customers well. He must not *overservice* them, giving them more trucks than they can handle adequately. He must make his decision based not on how many trucks a customer orders but on how many yards per hour he will need. Moreover, the dispatcher must not hesitate to reroute one or more of the trucks dispatched to a customer if the customer's delivery discharge rate is slower than promised.

When the customer of a producer's competitor calls in an order in the morning, the dispatcher should not promise delivery if it interrupts his planned schedule. The customer more than likely called his usual supplier too late to be serviced. Chances are he won't be heard from again until the same thing happens. Also, there is always a good chance that he is behind in his accounts receivable and has been told that there would be no further deliveries made to him until his back bill is taken care of. If he has not paid his usual supplier, he is unlikely to pay his new one.

The manufacture of concrete is a service-oriented business. Although there are occasional circumstances in which a late delivery cannot be avoided, the customer, whenever possible, should be notified of the delay and of the approximate time he can expect its arrival.

CONCRETE SALESMAN'S CUSTOMER SURVEY

It is a good practice to have the salesman, once or twice a year, ask pertinent questions on his visits to the customers. Along with his usual "Thank you for your business," he should ask about the service. Are the drivers cooperative? When he has occasion to call the plant office on a question of billing, does the customer receive prompt and courteous attention? Is the dispatcher friendly, courteous, and helpful?

A survey of this kind will always bring up a few gripes—some customers either are never satisfied or wouldn't admit it if they were. But it may well turn up a weak link in public relations or operational procedures that could then be strengthened.

THE PROJECT JOB CONTRACT

Lastly, the concrete producer may avoid many problems by including certain items in his contract on the larger projects. He might, for example, state the following:

1. There will be no additional charge for short loads or waiting time *unless* the practice becomes habitual or excessive.
2. There will be no charge for overtime after 4:00 P.M. *unless* a large delivery is requested for later than 8:00 A.M.

The two examples above give the concrete supplier some basis for complaint should the customer take too many costly advantages.

16 / HANDLING CONCRETE COMPLAINTS

Every manufacturer who produces a product for public distribution is bound to have an occasional complaint. But the ready-mix producer who delivers his product in all kinds of weather, for different applications, to customers with various degrees of talent is inevitably subjected to misunderstandings and grievances. There are more articles written on concrete problems, and their prevention and repair, than on any other topic in the construction industry.

SHORT (UNDER) YIELD

One of the frequent complaints, usually unjustified, is that of underyield. The customer complains he was shorted on a delivery by ½ yd or more. It is not always easy to satisfy his objections. Often, by checking the measurements of the concrete in place, the customer can be shown how he miscalculated the necessary yardage. But after the concrete has hardened, checking the depth of the concrete, for possible irregularities of the subgrade, is impossible.

If there is a batch recorder at the concrete plant, it can provide proof of the mix weights. A talk with the driver who made the delivery may be helpful if no such recorder was in operation. On the other hand, some producers will not honor a complaint of short yield unless the concrete is still plastic, permitting the subgrade to be measured.

Where the question is doubtful and the customer is not a habitual complainer, some credit adjustment may have to be made. In general, however, such an adjustment should be a last consideration. Salesmen are often too quick to supply the additional free concrete to pacify the irate customer.

THE JOB FOREMAN

It should be kept in mind that the person who registered the complaint may not have personally been on the job at the time of the concrete delivery and is acting on behalf of his job foreman, who complained to his boss to cover his own mistake. Some foremen and superintendents pride themselves on ordering just enough concrete to do the job with little or no concrete left over. When they shave their orders too thin, they must then order another yard to complete the job. The delay in finishing and the probable charge for the short-load delivery are far more costly than it would have been had they ordered a little extra concrete to compensate for subgrade irregularities or form expansion. In any case, many of these men, rather than admit their mistakes, will complain of short yield.

A lower air content than the mix was designed for is not a factor, in most cases, of lower yield. The 1 to 1½ percent of normally entrapped air in concrete will generally compensate. The complete absence of an air-entraining agent, when it should have been included, will reduce yield by a maximum of only 1 cu ft/cu yd. Agreed, on a large pour, this factor may exert some influence on yield but, normally, to a very minor degree.

LEFTOVER CONCRETE

Because of the strict ecology laws recently enacted, which greatly restrict the areas in which leftover concrete can be dumped, there is a greater tendency to reuse concrete returned to the plant. Although the concrete returned may often be perfectly satisfactory to the next customer, the practice has its drawbacks and its risks. There are few ready-mix drivers who can accurately tell how much yardage is left in their barrels. If a driver guesses wrong and the dispatcher gambles by adding sufficient new yardage to supply a second customer's requirements, he may well have to repeat the delivery with short yardage to complete the delivery. This practice must always be considered when a complaint of short yield is registered.

At best, using leftover concrete in hot weather is risky business. It may be done with reasonable success in cool weather, if the concrete has not been in the truck barrel too long, and if extra cement has been added to compensate. However, the temperature of the concrete may well have reduced the air content to a point where the resistance of that concrete to freeze/thaw cycles is diminished. It is far better to avoid the practice of using leftover concrete.

ANSWERING THE COMPLAINT

All customers of the concrete producer should be made aware that the responsibility of the supplier ends after the discharged delivery load has been accepted. Any slumps, concrete cylinders, or air tests to evaluate the quality of the concrete must be performed at the discharge end of the truck chute (ASTM C 94). However, there are exceptions to this rule. If the concrete fails to set properly, because of an excessive dosage of a set-retarding admixture, or for any other reason that could not have been determined at the time of delivery, the producer should be responsible for its removal and replacement, plus costs. Cracks, weathering, and similar problems should be assumed to be the fault of the normal behavior of concrete, or caused by poor workmanship.

Regardless of the cause of the concrete problem, it is the producer who is called on, most often, to defend the product. The producer's representative who usually answers the complaint should keep in mind that the complainer may, in fact, know the cause of the problem and really be calling for help and a cure, rather than be calling simply to blame the producer.

The producer should never go to the problem site with a chip on his shoulder, prepared only to defend his product. He should show no evidence of annoyance but rather manifest concern for the customer's problem. He should display understanding and sympathy and a sincere willingness to help in the solution of the problem. He should hear the customer out, get all the pertinent facts, and determine as best he can the source of the problem and the probable cure.

Take the case of a sidewalk, for instance. Let us assume that the concrete was placed in the fall of the year, and in the following spring it showed evidence of spalling near the curb line, but not further back. It could be pointed out to the customer that deicing salts, which were probably used on the road, were splashed onto the sidewalk by passing traffic. This explanation is certainly logical because of the fact that the concrete was placed in the fall of the previous year, since cool weather does not promote strength gain. It may also be pointed out that it is almost impossible to mix a load of concrete in such a way that only the section near the curb is faulty. It is diplomatic to avoid the customer's method of curing as a possible source of the problem, or his ability or workmanship. The old axiom of lose the argument but keep the customer is still good advice.

The classic complaint about badly pitted concrete on garage floors or slabs adjacent to gasoline pumps at service stations is easily addressed. Even when the customer with the problem contends no deicing salts or chemicals were used, any car parked on these areas carried these agents on the tires or undercarriage from the roads, which were so treated.

PRODUCER'S RESPONSIBILITY

How long should a concrete supplier be held responsible for his product? What is his obligation in the case of concrete that has performed well for years but that later shows signs of deterioration? It is, of course, true that good-quality concrete placed by experienced contractors should get stronger with age. But there can't be any guarantee that this will always happen. There are too many extrinsic factors that influence the performance of the concrete: heavy loads in excess of design, nearby drilling or blasting, erosion of the subgrade by heavy rains or floods, extreme heat that buckles pavement, the deterioration of expansion joints, or the filling of the expansion joints with foreign matter preventing the joints from performing their design function. Sometimes the probable cause of a concrete problem is obvious, while at other times it is impossible to explain the problem after the fact of its occurrence.

Even when the cause of the problem is obvious and clearly explained, the complainer may not acknowledge it. He may be, and often is, convinced that the recipient of the complaint, the producer or his representative, is talking over his head—or double-talking—to confuse him and to cover up a mistake. It is recommended that, when answering a complaint, the representative or salesman carry with him a supply of literature from a well-known concrete association to back his diagnosis and recommendations.

REGISTERING THE COMPLAINT

Any complaint about faulty concrete should be registered not by the owner of the concrete, unless he is a do-it-yourselfer, but by the contractor who placed the concrete. This contractor should be present on the problem site when the concrete supplier answers the complaint call. After studying the problem, the supplier should discuss the cause or solution with the contractor out of hearing of the owner.

It is unfair to expect the supplier to answer a complaint call directly to an owner without the contractor present. This puts the supplier in the awkward and compromising position of having to offer the owner an explanation for the problem while at the same time trying to protect the reputation of his customer, the contractor who placed the concrete. On the other hand, if the contractor hasn't shown up to answer the complaint, in order to avoid facing *his* problem, it would be worthwhile to ask a few questions of the owner. For example: Did it rain during or shortly after the placement of the concrete? What was the ambient temperature on the day the concrete was delivered? How many of the contractor's men were present? Did they leave the job after the concrete was placed and come back later to finish? Had the concrete been cured?

It is certainly unfair, but it is general practice to blame the concrete producer for problems that are created by faulty workmanship. It is, and probably will continue to be, the name of the game to seize on the producer as the scapegoat for all the mistakes that have been committed by others. The only possible solution to that tendency is for all concerned to learn as much as possible about concrete and the concrete industry.

However, mistakes by the producer can and do happen. In such situations the supplier should tell the truth and take the blame and the responsibility for correcting the error. On a short-yield problem it should be pointed out to the customer that the shortage was certainly not deliberate, as the cost of delivering the short load could well eliminate any profit made on the original order. The producer also takes every precaution to prevent botched batches for the same reason.

17 / GLOSSARY OF CONCRETE TERMS

Most of the concrete terms in the concrete industry are defined in this chapter. Many of these definitions were taken from the *Construction Dictionary* of the National Association of Women in Construction, who publish a dictionary of all terms used in construction, and used with their permission. Their book is an excellent reference source for anyone in the construction field.

Many of the terms used in this glossary are slang words, used in the jargon of the concrete trade.

Abrams's law The rule stating that the ratio of the amount of water in a mix to the amount of cement determines the strength of the concrete.

Abrasion resistance The ability of a surface to resist wear by friction.

Absolute volume The volume of solid particles excluding the space between them; volume without voids. (See Example 3-1 under "Absolute Volume" in Chapter 3 for formula.)

Absorption The increase in weight of a material due to moisture absorbed into the pores of its structure.

Accelerator An admixture which, when added to concrete, increases the rate of hydration and thus shortens the time of set.

Acid etching The use of an acid, usually muriatic, to improve bond on a surface to be repaired.

Additive Any material added to the cement or concrete mix other than the standard materials.

Admixture A material added to a concrete mix to produce a desired quality feature.

Aggregate Any of several materials such as sand, stone, slag, or expanded shale used to make concrete.

Aggregate/cement ratio The weight of the aggregate divided by the weight of cement.

Agitation The mixing of concrete in a gentle turning of the mixer to maintain plasticity.

Air entrainment The introduction of an air-entraining agent to cement or concrete to produce billions of air bubbles in a concrete mix to improve workability and frost resistance.

Air meter An instrument used to measure the air content of concrete.

Air void A pocket of air entrapped in concrete.

Algae A form of plant life found in water; a pond scum.

Alkali aggregate An aggregate that reacts chemically with the alkalies in portland cement.

Alkaline salts Present in some soils. Water can dilute these salts and penetrate into the concrete with destructive effects.

Alumina Aluminum oxide (Al_2O_3).

Architect A licensed person whose profession is to design structures, draw up plans, specify materials, handle the bidding, and generally supervise the construction.

Architect-engineer An individual or firm that offers the professional services of both architect and engineer.

Backfill Material, such as earth or sand, used to replace material removed during construction.

Bag (sack) A quantity of portland cement (94 lb).

Ball test A test to determine the slump of freshly mixed concrete, using a Kelly ball.

Bank run A gravel with a maximum-size aggregate obtained from a natural deposit without careful gradation; a combination of sand and coarse aggregate.

Bar A piece of metal used to reinforce concrete.

Bar chair An instrument used to support reinforcement bars during the placement of concrete.

Barrel A weight measure of portland cement, consisting of 4 bags (376 lb).

Batch A quantity of concrete mixed at one time.

Batcher The individual who weighs out the materials for a concrete mix.

Batch mixer A circular metal machine that mixes a batch of concrete.

Below grade Below ground level; usually used to refer to a wall or floor.

Bituminous concrete A material composed of graded fine and coarse aggregate and bound together with asphaltic cement.

Bleeding The appearance of water on the surface of concrete due to the settlement of solids in the mix.

Blisters The irregular raising of a thin layer of cement paste on the concrete surface during or shortly after the finishing operation.

Boiled linseed oil Linseed oil in which materials have been added to make the oil harden more rapidly.

Bond Adhesion of concrete to reinforcement or to other surfaces against which it is placed.

Bonding agent In concrete, that material used between old and new concrete to ensure adhesion.

Bridge deck The load-bearing floor of a bridge which carries and spreads the load to the main beams.

Buggy A two-wheeled or motor-driven cart used to transport small quantities of concrete from hoppers or mixers to forms.

Bulk cement Cement which is delivered in large quantities rather than in bags.

Bull float A tool used by masons to smooth and level concrete floors after screeding.

Burlap A coarse fabric of jute or hemp used in covering finished concrete for water-curing.

Bush hammer A hammer with a serrated face which is used to obtain an exposed-aggregate finish.

Calcium chloride An admixture used to accelerate the set of concrete ($CaCl_2$).

Calcium hydroxide Slaked lime ($CaOH_2$).

Capacity In truck mixers, the maximum volume of concrete to be mixed or carried in a particular mixer or agitator.

Carbon black A finely divided carbon produced by the incomplete burning of oil or gas; used to color concrete black.

Cast-in-place Concrete which is delivered in the plastic state and left to harden as part of a structure, as opposed to precast concrete.

Cavitation The pitting of concrete, resulting in the appearance of honeycombed concrete, caused by the collapse of bubbles in water flowing by obstructions.

Cellular concrete A lightweight concrete made of portland cement with lime/silica or lime/pozzolan pastes.

Cement A material made by burning and grinding a mixture of clay and limestone with the addition of chemicals to control time of set.

Cement/aggregate ratio The ratio by weight or volume of cement to aggregate.

Cement factor The number of bags or pounds per cubic yard. One bag (94 lb) equals 1 cu ft.

Cement hydration The reaction which takes place between cement and water.

Cement, hydraulic A special cement capable of hardening under water.

Cement, low-alkali A portland cement containing a relatively small amount of sodium and/or potassium.

Cement, natural A product obtained by pulverizing calcined limestone burned at a temperature high enough to drive off carbon dioxide; sometimes used in conjunction with portland cement.

Cement, neat A mixture of cement and water without other materials added, i.e., cement paste.

Central mix plant A plant that mixes the batched concrete weights into a plastic state for delivery to a jobsite, as opposed to a transit mix plant, which mixes the concrete on the jobsite.

Chemical resistance The degree to which concrete, or special coatings or floor toppings, can resist chemical attack.

Chert A very fine grained tough rock composed mainly of silica, commonly found in limestone beds.

Clamshell A crane bucket with two jaws which clamp together by their own weight when lifted by a closing line. It is used in handling aggregates.

Clinker A partially fused product of a kiln which is ground to make cement.

Coating Material applied to a surface by brushing, rolling, spraying, or troweling, to preserve, protect, or seal.

Coefficient A factor that contributes to produce a result.

Coefficient of variation The standard deviation divided by the mean.

Cofferdam A watertight enclosure within which excavation is done for a foundation.

Compaction The process of inducing closer packing of the solid particles in plastic concrete by vibration or tamping.

Compressive strength The measured maximum resistance of a concrete specimen to axial loading, expressed in pounds per square inch (psi).

Concrete A combination of portland cement, fine and coarse aggregates, and water, designed to produce a mix of specified strength.

Concrete mixer A metal drum of various sizes with special blades inside to mix the ingredients of concrete.

Concrete pump A machine which forces concrete through a pipeline or hose to a placing site.

Consistency The ability of freshly mixed concrete to flow; slump.

Contractor A person or company who agrees to supply labor and materials to do work for a certain price under the terms of the contract.

Core A cylindrical sample of hardened concrete obtained by use of a drill specially made for that purpose.

Core test A procedure for evaluating compressive strength, based on a concrete sample taken from hardened concrete.

Corrosion The deterioration or disintegration of concrete.

Corrugated Having alternate ridges and valleys in parallel.

Crane A mobile or stationary machine used for lifting and moving loads.

Craze cracks Numerous fine shrinkage cracks appearing on a concrete surface caused generally by the evaporation of surface moisture on windy days of low humidity.

Creep The deformation of concrete when subjected to sustained loads over a period of time.

Crusher A machine that reduces large-size aggregates into smaller sizes.

Curing Protective measures to prevent the escape of water from freshly placed concrete through evaporation.

Curling The warping of a flat slab, usually caused by the more rapid shrinking of the top of the concrete than the bottom.

Cylinder A mold for casting samples of fresh concrete for test in compression at a later period.

Darby A hand tool; a long, flat, rectangular piece of wood, aluminum, or magnesium, from 3 to 4 in. wide with a handle on top. It is used to float the surface of concrete immediately after screeding to eliminate high or low spots in preparation for the finishing operation.

Dehydration The loss of water necessary for the proper curing of concrete.

Deviation A departure from standard.

Diamond saw A circular saw set with diamonds in its cutting face and used in concrete to cut joints.

Discoloration A departure of color from that which is normal.

Dispersing agent An admixture that breaks up cement flocculants in concrete, reducing water demand for a given slump and adding plasticity to the mix.

Dolomite A rock containing equal amounts of calcium and magnesium carbonates; used as coarse aggregate in concrete.

Dry mix All the ingredients for a concrete mix without the presence of moisture.

Dusting The appearance of a fine white powder on the surface of concrete.

Dust on The practice of casting dry cement on the concrete surface to absorb bleed water.

Edger A curved-end hand tool used to produce a rounded corner at the edge of a concrete slab.

Efflorescence A deposit of soluble salts appearing on the surface of concrete, generally created by the penetration of water through the concrete.

Elasticity The ability to recover the original shape and size after deformation.

Elastic limit The limit of stress from which the concrete cannot fully recover.

Elastic sealants Used when concrete cracks are expected to remain active.

Elastomers Plastic, synthetic, or rubber concrete additives to increase bond strength and add other flexible properties.

Elephant trunk A sectional metal tube for placing concrete where segregation may occur in high forms.

Entrapped air Air present in concrete that is not added purposely as by an air-entraining agent.

Epoxies Specially formulated resins used for paints, coatings, and mortars in new construction and in the repair of faulty concrete. They are sold in two-component systems, and when mixed, they produce outstanding quality features.

Expanded shale A lightweight aggregate for concrete.

Expansion Enlargement of concrete caused by temperature rise, absorption of water, and freezing.

Expansion cement A special cement that expands as it sets to reduce normal plastic shrinkage.

Expansion joint A separation between adjoining sections of concrete to allow for expansion.

Expansion joint filler A flexible material to prevent foreign matter from entering a joint.

Exposed aggregate A concrete finish achieved by removing the surface mortar and exposing the aggregate.

External vibration A method of vibrating concrete by attaching vibrators on the outside of the forms.

False set A premature stiffening of concrete that can be made plastic by vibration.

Fat mix A concrete mix containing a high cement factor.

Field-cured Cured on the construction site rather than in a laboratory curing room; used to refer to concrete cylinders used for testing.

Final set When concrete has hardened enough to resist penetration of a weighted test needle.

Fine aggregate That portion of aggregate that passes the ¼-in. screen.

Fineness modulus (FM) A factor obtained by adding the retained cumulative weights of aggregate on specified sieves and dividing the total by 100. For fine aggregate the sieves used are ⅜ in., No. 4, No. 8, No. 16, No. 30, No. 50, and No. 100.

Finish The appearance of a concrete surface after the finishing operations have been completed.

Finisher A skilled mason experienced in troweling concrete.

Finishing Compacting and leveling concrete surfaces to produce a desired appearance.

Finishing machine A power-operated bladed machine used on the surface of semihardened concrete.

Flash set A rapid hardening of plastic concrete that resists methods to return it to a plastic state.

Flexural strength A factor of resistance measuring the extent to which a concrete beam will bend before it breaks.

Floating foundation A type of foundation made to carry the weight of a building to be erected on unstable soil.

Flow A measure of the consistency of freshly mixed concrete.

Fluosilicates Salts of magnesium or zinc used on concrete surfaces as hardening agents.

Fly ash A pozzolon; a finely divided residue obtained from coal-burning powerhouses used in concrete in addition to or as a replacement for a percentage of the cement.

Fog curing Storage of concrete test samples in a moist room, under controlled temperatures, kept damp by a fine fog-like spray.

General contractor The prime or main contractor responsible for the overall work.

Georgia buggy A wheelbarrow to carry concrete to the place of deposit.

Gillmore needle An instrument used to determine the time of initial and final set of concrete.

Go-devil A fabricated device, or rolled-up burlap, put through the end of a concrete pump line and forced through the pump line with water, for cleaning the line after its use.

Gradation Sieve analysis of aggregates to test for particle size specification conformation.

Granolithic concrete Concrete made of selected aggregates of extra hardness to improve surface wear.

Gravel Granular aggregate with a usually smooth-faced appearance used as the coarse material in concrete mixes.

Grinding aids Materials used in the manufacture of cement to disperse the finely ground product.

Grooved joint A construction joint created by forming a groove in the surface of concrete to control cracks.

Grout A mixture of cement, sand, and water.

Guniting A method of spraying cement, sand, and water through a special gun under pneumatic pressure.

Gypsum An important mineral in the production of cement in controlling time of set.

Hardener A chemical applied to concrete surfaces to reduce wear and/or cure dusting. In a two-component epoxy the hardener is the lesser amount.

Haul distance The distance material is carried.

Haydite An expanded clay used as aggregate for lightweight concrete.

Heat of hydration Heat produced by the chemicals in cement reacting with water.

High-density concrete Concrete of very heavy unit weight obtained by using heavyweight aggregate; generally used for radiation shielding.

High early cement A finely powdered portland cement with a chemical composition that induces quick setting.

High early concrete Any concrete capable of attaining higher strength at an earlier age than normal concrete.

Honeycombing Unintentional exposure of the coarse aggregate in a section of hardened concrete, often caused by using excessive coarse aggregate, leakage of mortar through the form, low-slump concrete, or improper or insufficient vibration.

Hose coupling A connection between two hoses or between a hose and a steel pipe.

Hot cement Cement of high temperature due to insufficient cooling after manufacture.

Hydraulic cement A type of cement which hardens under water.

Hydrometer An instrument, usually a tubelike floating glass, used to measure the specific gravity of fluids.

Impervious Resistant to water penetration.

Initial set That degree of fresh concrete hardening enough to permit foot traffic to permit the finishing operation to begin.

Inspection The examination of concrete work before, during, and after completion for compliance with contract specifications.

Integral waterproofing A method of waterproofing concrete by adding a specific admixture to the mix.

Internal vibration Compaction of concrete by introducing a vibrator into freshly placed concrete.

Jobsite The location of a construction project.

Joint sealant Material used to prevent the entry of water or foreign materials into concrete joints.

Kelly ball An instrument used for determining the slump of concrete.

Keyway A recess or groove on top of concrete which is to receive an additional course.

Laboratory A material's testing laboratory retained by the architect or engineer for testing and/or inspection of concrete.

Laitance A layer of objectionable weak mortar on the surface of concrete, caused by excessive bleeding and/or overtroweling.

Latex emulsions Products of water and rubber or synthetic resins generally used for repairing faulty concrete.

Lean concrete A concrete mix with a low cement content.

Lightweight concrete Concrete designed with special aggregates to greatly reduce the unit weight of normal concrete.

Loam A soil composed of clay, sand, silt, and organic matter.

Low-alkali cement A portland cement containing a relatively small amount of sodium and/or potassium.

Magnetite A heavy material with a specific gravity of 5.0 plus, used as aggregate in high-density concrete.

Mason A workman skilled in finishing concrete.

Masonry cement A special hydraulic cement manufactured for use in mortars; used in erecting block, brick, etc., where greater plasticity is required.

Mass concrete Any large volume of cast-in-place concrete.

Matrix The cement/sand/water paste which bonds the coarse-aggregate particles.

Membrane curing A method of curing concrete after finishing by liquid-curing compounds or with a covering to prevent evaporation of water from the concrete.

Mixer A barrel-shaped machine used for mixing the materials of concrete; a ready-mix delivery truck.

Mixing time The period of time necessary to mix concrete ingredients into a plastic state; often measured by the number of total revolutions of the mixer.

Modulus of elasticity The change of stress with respect to deformation. It is the measure of resistance of a material to deformation.

Mohr's scale A scale used to indicate relative hardness in ascending degrees of from 1 to 10. A diamond is rated at 10.

Moisture content The percentage of water in wet aggregates, not including the absorption factor.

Monolithic concrete Concrete placed or cast with no joints other than construction joints.

Monolithic surfaces A dry mixture of various compositions sprinkled on a moist fresh concrete surface and worked into the concrete by floating.

Mortar A mixture of cement, water, and sand.

Mud A slang expression for mortar.

Mud jacking Elevating a sunken concrete slab by forcing a slurry through a hole bored in the surface of the slab.

Mudslab A layer of concrete below a structural slab or footing over an uncertain subgrade.

Natural cement A product obtained by pulverizing calcined limestone burned at a temperature high enough to drive off carbon dioxide; sometimes used in conjunction with portland cement.

Neat cement A mixture of cement and water.

No-fines concrete A concrete mix containing cement, water, and coarse aggregate.

Nontilt mixer A stationary horizontal drum-type mixer with rotating blades; occasionally used in ready-mix plants but in general use in concrete block and pipe plants.

No-slump concrete Concrete with a slump of 1 in. or less.

Ottawa sand A silica sand of naturally rounded particles mined in Ottawa, Illinois; used in the making of mortar specimens for testing cement.

Oversanded Containing a higher proportion of sand than necessary for good workability of a concrete mix.

Overvibration Excessive use of vibrators in placing fresh concrete that brings an excess of fines to the surface.

Pea gravel That size gravel which passes the ⅜-in. sieve and is retained on the No. 4 sieve.

Percent fines The percentage of material passing the No. 200 sieve in a sand analysis test.

Perlite A very lightweight aggregate used in concrete as an insulating material.

Permeability The lack of ability to restrain the passage of water.

Pit-run gravel Gravel direct from a pit; also referred to as *run-of-bank* (ROB).

Placing The depositing of freshly mixed concrete in the place where it is left to harden; often erroneously referred to as *pouring.*

Plain concrete 1. Concrete without the addition of any admixture. 2. Concrete without reinforcement.

Plastic cracking Cracking that occurs on the surface of concrete after it is placed and while it is still plastic.

Plasticity A characteristic of fresh concrete that makes it homogeneous and readily workable.

Plasticizer An admixture added to a concrete mix to increase plasticity; an aid in improving pumpability.

Polymers Lignosulfonates of high molecular weight.

Ponding A method of water curing flat concrete surfaces by placing a small earth dam around the perimeter.

Popout A shallow conical depression in concrete surfaces caused by the expansion of unsound aggregates.

Posttensioning A method of prestressing reinforced concrete after the concrete has hardened.

Potable water Water that is acceptable for drinking.

Pour A slang word for "place."

Pozzolans Siliceous materials in finely divided form which, in the presence of moisture, chemically react with calcium hydroxide to form compounds possessing cementitious properties. Examples: fly ash, volcanic ash, calcined clays and shales, and ground limestone.

Precast Cast and cured in other than its final position.

Preplaced-aggregate concrete Concrete produced by filling a form with coarse aggregate and injecting grout to fill the voids.

Preshrunk concrete Concrete that has been mixed for an hour or more before placing to reduce shrinkage.

Prestressed concrete Concrete placed in special forms with tightly drawn steel wire whose tensions are released after the concrete has hardened to increase flexural strength.

Proportioning The designing of a concrete mix with the proper percentages of cement, water, and aggregates for a required strength and workability.

Pycnometer A bottle device used in testing for the specific gravity of an aggregate.

Quarry A rock pit.

Quartering A method of obtaining a representative sample of aggregate for test.

Quartz A crystalline silica; the major portion of sand, gravel, or sandstone.

Rake finish A method of finishing concrete by creating deep grooves with a rake on the surface of concrete which is to receive a special topping or second layer.

Ready-mixed concrete Concrete produced for delivery to a purchaser in a plastic state from either a central mix or transit mix plant.

Rebar A bar of reinforcing steel.

Refractory aggregate A material used in concrete to offer resistance to high temperatures; concrete with low thermal conductivity.

Reinforced concrete Concrete containing reinforcing steel to improve resistance to load, stress, and other forces.

Retardation A delay of concrete set beyond the normal time required.

Retarder An admixture added to the concrete mix to delay the normal set time.

Retempering The addition of water to a mix which has started to set and caused slump loss. The total water added, rather than the original amount used, will determine the strength of the concrete.

Revibration The vibration of concrete already compacted but before initial set.

Revolution The turning of a mixing barrel one complete cycle.

Ribbon loading A method of loading concrete into a mixer after batching in which all ingredients are fed simultaneously.

Rodability The ability of fresh concrete to be compacted by means of a tamping rod.

Rotary A rotating machine.

Rotary float A motor-driven revolving disk with trowel blades for finishing concrete floors.

Rubbing The rubbing down of concrete immediately after form removal, with burlap and a water/cement paste, or with a Carborundum brick.

Sack (bag) A quantity of portland cement (94 lb).

Sand *See* Fine aggregate.

Sand/aggregate (S/A) ratio The ratio of fine aggregate to coarse aggregate in a concrete mix, by weight or by volume.

Sand streaking The appearance of sandy surface streaks on a concrete wall caused by excessive bleed water rising to the top between the face of the concrete and the form.

Saturated Unable to absorb more water.

Saturated surface dry (SSD) Moisture-saturated internally but without free water externally. All concrete mixes are designed for aggregates in this condition.

Sawdust concrete Concrete with sawdust as the main ingredient for nailing.

Sawed joint A joint cut in hardened concrete of an inch or more with a diamond-tipped disk on a power-operated circular saw.

Scabbler A machine used to remove loose surface concrete in preparation for resurfacing.

Scaling (spalling) The development of concrete surface deterioration causing unwanted exposure of the coarse aggregate.

Scarifier A machine to remove loose concrete from a faulty concrete surface.

Score To scarify or roughen a concrete surface to assure better bond for the repair material.

Screed A stiff straight-edged tool usually of magnesium for striking off concrete surfaces.

Screen A sieve, for gradation of concrete aggregates.

Segregation The lack of ability for concrete to remain homogeneous, permitting the coarse aggregate to separate from the mix; usually associated with concretes of high slump.

Separation The separation of fine and coarse aggregation in handling or stock piling material in conical piles.

Set Of concrete, when the cement paste has lost its plasticity. 1. Initial set refers to the first stiffening. 2. Final set is reached when the concrete is completely hardened.

Setting shrinkage A decrease in volume due to concrete aggregates, having the highest specific gravity in a mix, settling in the mass.

Shear modulus The modulus of rigidity.

Short yield A concrete mix of less than 27 cu ft.

Shotcreting Forcing mortar or concrete through a hose, under high pressure; generally used to repair faulty vertical concrete surfaces.

Shrinkage A reduction in volume of a concrete mass; usually referred to as *drying shrinkage.*

Shrinkage crack A crack caused by the shrinkage of plastic concrete as it sets.

Sieve number A number referring to the number of openings per square inch in fine-aggregate sieves or screens.

Silt The fine material in sand passing the No. 200 sieve.

Site The location of a construction project.

Slab A concrete floor.

Slag The leftover product from steel manufacturing.

Slag concrete A concrete mix using slag as the aggregate.

Slipform A form which is pulled or raised as concrete of low slump is placed.

Slump A measure of consistency of freshly mixed concrete.

Slump cone A device to measure slump. The wetter the concrete, the higher the slump.

Slump test The procedure used to measure slump.

Slurry A mixture of cement, sand, and water.

Slushing A mixture of cement and water mixed to a slushy consistency and used as a bonding agent between old and new concrete.

Soft particle A particle of aggregate that fails to meet a soundness aggregate test.

Solvent A liquid that dissolves or can dissolve another substance.

Soundness The ability of an aggregate to resist weathering and/or wear and to react favorably with the alkalis in cement.

Soup Slang for overly wet concrete.

Spalling (scaling) The development of concrete surface deterioration causing unwanted exposure of the coarse aggregate.

Specific gravity The bulk specific gravity of aggregates is defined as the ratio of the weight in air of a given volume of material to the weight in air of an equal volume of water.

Specific surface The surface area of a specific amount of cement; used to express the particle fineness.

Straightedge (screed) A rigid straight piece of wood or metal to strike off a concrete surface.

Striking off Removing excess concrete on a concrete slab with a straightedge to level the surface.

Stripping Removing forms after the placed concrete has hardened.

Stucco A plastic mortar troweled on an exterior exposed surface with a desired texture; different from podging, which is a flat finish.

Sulfate resistance Ability of concrete to resist sulfate attack.

Supplier A concrete or aggregate producer.

Surface area In aggregate, the smaller the particle size, the greater the surface area. The lower the fineness modulus (FM), the greater the surface area.

Surface moisture The free water retained on the surface of aggregates.

Surface voids Cavities in a concrete wall usually caused by entrapped air.

Tailings Waste material separated in the production operation of stone or gravel.

Tamping Compacting fresh concrete by rodding.

Temperature rise The increase of concrete temperature caused by the hydration of cement and water, additionally influenced by hot weather, hot cement, and warm water and aggregates.

Tensile strength The ability of concrete to resist downward pressure before breaking. Concrete without reinforcement has low tensile strength (approximately 600 psi).

Terrazzo A type of concrete floor made of small pieces of marble or colored stone embedded in the surface and finished with a high polish.

Tie rod A steel rod used to keep wall forms from spreading under the pressure of freshly placed concrete.

Tilt-up concrete Concrete that is cast horizontally on a jobsite and raised into place after removal from forms.

Ton 2000 lb. A cubic yard of concrete weighs approximately 2 tons.

Topping A surface layer of concrete, mortar, epoxy, etc., over new or faulty concrete surfaces.

Traprock A natural rock used as aggregate for concrete.

Tremie A pipe through which concrete can be placed under water.

Tremie concrete A specially designed concrete placed by means of a tremie.

Trowel A flat-bladed steel or magnesium tool used by hand in the concrete finishing operation.

Troweling machine A power-operated device with three trowel blades rotating in a horizontal motion for finishing concrete floors.

Truck mixer A mixer mounted on a truck for concrete delivery.

Turbidimeter A device used for measuring the particle-size distribution of cement.

Undersanded Having insufficient fine aggregate for good workability.

Unit water content The number of pounds or gallons of water per cubic yard of concrete. In the water/cement (W/C) ratio, it is the number of pounds or gallons of water per bag of cement.

Vacuum concrete Concrete from which water is extracted by means of a vacuum mat.

Vermiculite A very lightweight aggregate used in concrete for insulating qualities.

Vibration limit The time when concrete has set hard enough not to be returned to a plastic state by vibration.

Vibrator An oscillating hand tool or machine for compacting concrete.

Voids Unoccupied spaces between particles of aggregates; entrapped air pockets in concrete.

Volume change An increase or decrease in volume. Concrete shrinks when it sets. It expands when wet and shrinks when it dries. It expands when it freezes and shrinks when it thaws. It expands when it is hot and shrinks when it is cold.

Warehouse set The partial set of cement stored over a period of time exposed to moisture.

Wash water That water used for washing out a concrete truck mixer after each delivery, obtained from a water tank mounted on the truck.

Water/cement (W/C) ratio The number of pounds or gallons of water used per bag of cement in a concrete mix; determines the strength of the hardened concrete.

Waterproof cement A cement made with a water repellent to make it watertight.

Water-reducing agent An admixture used in concrete to decrease the normal water demand for a given slump.

Water-repellent cement A hydraulic cement with a water-repellent agent added to resist absorption of water; generally used to repair leaks in concrete basements.

Water stop Thin sheets of rubber, plastic, or other material inserted in a construction joint to obstruct the seepage of water through the joint.

Weigh batcher The person responsible for weighing concrete ingredients for concrete in a batch plant.

Well-graded aggregate Aggregate that has particle-size distribution conforming to gradation specifications to produce maximum density.

Workability That characteristic of concrete that produces a mix favorable to ease of placing and finishing.

Yard A cubic yard of concrete (27 cu ft).

INDEX

About the Author

John C. Ropke has been in the concrete industry for nearly 40 years. Currently an independent concrete consultant, he has been a concrete inspector in the construction of dry docks and airports, an inspector trainer, a quality-control engineer, and a concrete troubleshooter—and has worked for the U.S. Navy, the U.S. Testing Company of New Jersey, the Johns-Manville Corporation, and other firms. During the past 20 years he has conducted a great many popular and well-attended seminars on concrete and concrete problems.